AAPS Advances in the Pharmaceutical Sciences Series

Volume 21

The AAPS Advances in the Pharmaceutical Sciences Series, published in partnership with the American Association of Pharmaceutical Scientists, is designed to deliver volumes authored by opinion leaders and authorities from around the globe, addressing innovations in drug research and development, and best practice for scientists and industry professionals in the pharma and biotech industries.

More information about this series at http://www.springer.com/series/8825

Russell Weiner · Marian Kelley
Editors

Translating Molecular Biomarkers into Clinical Assays

Techniques and Applications

Springer

Editors
Russell Weiner
Translational Biomarkers and Personalized Medicine
Edison, NJ
USA

Marian Kelley
MKelley Consulting LLC
West Chester, PA
USA

ISSN 2210-7371 ISSN 2210-738X (electronic)
AAPS Advances in the Pharmaceutical Sciences Series
ISBN 978-3-319-40792-0 ISBN 978-3-319-40793-7 (eBook)
DOI 10.1007/978-3-319-40793-7

Library of Congress Control Number: 2016943065

Printed on acid-free paper

This Springer imprint is published by Springer Nature
The registered company is Springer International Publishing AG Switzerland

The original version of the bookfrontmatter was revised: The volume number of the book was corrected. The Erratum to the bookfrontmatter is available at 10.1007/978-3-319-40793-7_19

Preface

The science of drug development is an evolutionary one, something we can see clearly in the world of biomarkers. Not so long ago the concept of using biomarkers in drug development was relegated to a few "research-only" kits. Bioanalysis in support of drug development focused solely on the pharmacokinetics (PK) of new drug entities. Unfortunately, PK-only approaches can no longer support today's drug development targets. While drug pipelines were full, and research was thriving, very few new drugs were being approved. To address the high attrition of drug development projects and attempt to improve the success rates, the FDA implemented the Critical Path Initiative in 2004. The FDA's strategy was to drive innovation by using the newest tools to more successfully translate discovery into viable therapies. Integrating biomarkers into the drug development process, especially in the pre-phase 3 stage, was central to the initiative. The challenges of including biomarkers in drug development were daunting until insightful colleagues started breaking it down to usable building blocks. A groundbreaking paper authored by Dr. Jean Lee and associates (2006) cited the need to improve the efficiency and economy of drug development by the use of well-validated biomarkers and biomarker assays. Workshops were developed, many notably led by Dr. Ron Bowsher, to help researchers understand the difference between a well-developed biomarker and a well-developed biomarker assay. This concept was solidified by Dr. Woodcock (2009) in a pivotal white paper which re-emphasized the need for an iterative method development process following the progress of the New Drug Entity through the drug development continuum and by the publication in 2011 by the ICH of the E16 Guidance for the Industry describing the requirements for biomarkers used in regulated submissions. Biomarker research has become such an integral facet of drug development that some pharmaceutical companies have implemented a policy of requiring a biomarker (target engagement, pharmacodynamics, and patient selection) to be included in all programs as a gating item to move into clinical development. The world of biomarkers has exploded in both depth and breadth. A cursory look over the table of contents in this e-book provides a thumbnail sketch of the innovation taking place in the modern day bioanalytical lab to help produce new drug entities quickly and efficiently. Finally, to illustrate the importance of using biomarkers to drive clinical development, the FDA issued a draft bioanalytical method validation guidance in 2013 that will now contain a biomarker section describing the "fit-for-purpose" need to validate biomarker methods when using the data to support a regulatory submission.

Edison, USA Russell Weiner
West Chester, USA Marian Kelley

Contents

Editors and Contributors

About the Editors

Russell Weiner is a senior level biopharmaceutical professional with expertise in pharmaceutical and biologic drug development, clinical trial conduct, regulated bioanalysis, use of biomarkers in early clinical decision making and developing companion diagnostic strategies. He is the Executive Director and head of Clinical Bioanalytical, Biomarkers and Companion Diagnostics team at Daiichi Sankyo Inc. He acquired his Ph.D. at the Albany Medical College. Dr. Weiner led clinical development teams responsible for transitioning therapeutics from discovery through phase two 'proof-of-concept' studies. With over 40 publications, he is also part of the Foundation of the NIH Cancer Steering Committee for The Biomarkers Consortium.

Marian Kelley is a leading expert in biotherapeutic drug development through biomarker application and assay analysis. She is the President and Founder of MKelley Consulting, LLC, a company that provides bioanalytical support in drug development through the validation of biomarker methods, immunogenicity studies, and cell-based assays. Her areas of expertise include immunogenicity assay screening, pharmacokinetics, and immunology. She was previously the Chair of the Ligand Binding Assay Bioanalytical Focus Group Steering Committee for the AAPS. Kelley is coauthor or coeditor of over a number of publications. She is also an extensively experienced trainer in validation of methods and quality control.

Contributors

Ronald R. Bowsher B2S Consulting, Indianapolis, IN, USA; B2S Consulting and AIT Bioscience, Franklin, IN, USA

Ken C.N. Chang Translational Molecular Biomarkers, Merck & Co., Rahway, NJ, USA

R. Aleks Davis B2S Life Sciences, Franklin, IN, USA

Marisa Dolled-Filhart Translational Biomakers, Merck & Co., Rahway, NJ, USA; Translational Medicine, Merck & Co., Rahway, NJ, USA

K. Emancipator Translational Biomakers, Merck & Co., Rahway, NJ, USA

Scott T. Fountain Pharmacokinetics, Dynamics and Metabolism—New Biological Entities, Pfizer Worldwide Research and Development, San Diego, CA, USA; Encinitas, CA, USA

Omar F. Laterza Translational Molecular Biomarkers, Merck Research Laboratories, Rahway, NJ, USA

Jean W. Lee BioQualQuan, Camarillo, CA, USA

Mingxiang Lin Analytical Research & Development, Merck & Co., Rahway, NJ, USA

Virginia Litwin Department of Hematology/Flow Cytometry, Covance Inc., Indianapolis, IN, USA

Matthew J. Marton Translational Molecular Biomarkers, Merck & Co., Rahway, NJ, USA

Andrew P. Mayer GlaxoSmithKline Pharmaceuticals Research and Development, King of Prussia, PA, USA

Hendrik Neubert Biomedicine Design, Pfizer Worldwide Research and Development, Andover, MA, USA

William Nowatzke Radix BioSolutions, Georgetown, TX, USA

Peter J. O'Brien Department of Pharmacokinetics, Dynamics and Metabolism, Pfizer, San Diego, CA, USA

Uma Prabhakar BRIM Biotechnology Inc, Taipei, Taiwan, People's Republic of China

Chad Ray Pfizer, San Diego, CA, USA

Manjula P. Reddy Oncology Translational Research, Janssen Research and Development, Janssen Pharmaceutical Companies of Johnson and Johnson, LLC, Spring House, PA, USA

Paul W. Rhyne Immunoassay Services, Q^2 Solutions, Marietta, GA, USA

Amy Kate Sasser Oncology Translational Research, Janssen Research and Development, Janssen Pharmaceutical Companies of Johnson and Johnson, LLC, Spring House, PA, USA

Usha Singh Translational Medicine, Merck & Co., Rahway, NJ, USA; Translational Biomakers, Merck & Co., Rahway, NJ, USA

Michael P. Sullivan Department of Bioanalytical Sciences, Worldwide Clinical Trials, Austin, TX, USA

Michelle Tseng Amerimmune Immunology Laboratory, O & O Alpan, LLC, Fairfax, VA, USA

Jeffrey A. Tsou Department of Reproductive and Genetic Health, Illumina, San Diego, CA, USA

Paolo Vicini Pharmacokinetics, Dynamics and Metabolism—New Biological Entities, Pfizer Worldwide Research and Development, San Diego, CA, USA; MedImmune, Cambridge, UK

Richard Wnek Translational Molecular Biomarkers, Merck & Co., Inc., Rahway, USA

Dianna Wu Translational Biomakers, Merck & Co., Rahway, NJ, USA; Oncology Translational Research, Janssen Research and Development, Spring House, USA; Translational Medicine, Merck & Co., Rahway, NJ, USA

Tim Wyant Department of Translational Medicine, Curis Inc., Lexington, MA, USA

Xuemei Zhao Translational Molecular Biomarkers, Merck Research Laboratories, Rahway, NJ, USA

Fit-for-Purpose Validation

Chad Ray

Abstract This chapter serves as an introduction to the terms and definitions of biomarkers as well as the validation assay process which includes: a standard curve, validation samples, accuracy and precision (reproducibility), limits of quantification, parallelism, dilution linearity, specificity and interference, stability, and normal range. Different biomarkers exist for the various stages of the drug discovery and development process. The main objective of this chapter is to introduce concepts that will facilitate the successful use of biomarkers in drug development—from selection, to validation, and then implementation.

Keywords Parallelism · Pharmacodynamics (PD) endpoints · Therapeutic efficacy · Definitive quantitative assay · Stability · Specificity · Accuracy

Biomarker Terms and Definitions

The application of biomarkers in drug development to support safety and efficacy testing is not a new concept, and in the early 2000s formal definitions were developed. The NIH consensus definition described a biomarker as a measurable endpoint that defines or relates to a biological, pathological, or drug activity process [1]. Shortly after those definitions were developed, the United States Food and Drug Administration released the Critical Path Initiative that outlined ways to increase innovation and enhance the probability of developing safer and more efficacious therapies [2]. Within that document, the FDA provides rationale for using biomarkers to increase the probability of technical success by treating the correct patients. Biomarkers are used throughout the drug discovery process from preclinical target engagement to late stage clinical efficacy assessment. There are multiple biomarker classifications that are described in Table 1. Biomarkers that demonstrate a pharmacological response to the therapeutic intervention are referred

C. Ray (✉)
Pfizer, 10628 Science Center Drive, San Diego 92122, CA, USA
e-mail: chad.a.ray@pfizer.com; Chad.Ray@iconplc.com

R. Weiner and M. Kelley (eds.), *Translating Molecular Biomarkers into Clinical Assays*, AAPS Advances in the Pharmaceutical Sciences Series 21,
DOI 10.1007/978-3-319-40793-7_1

Table 1 Biomarker classification

Attribute	Pharmacodynamic	Efficacy	Diagnostic	Prognostic	Predictive
Purpose	Demonstrate relationship to compound exposure	Demonstrate differences between compound treated and untreated subjects	Disease identification	Disease progression	Stratifies patients that will likely demonstrate efficacy on therapy
Stage of development	Preclinical and Phase I/II	Phase Ib/III	Phase III and safety testing	Typically used in treatment setting	Phase II/III
Regulatory	GLP/GCP	GLP/GCP	CLIA/CAP	CLIA/CAP	GLP until PMA or 510 K approval achieved

to as **pharmacodynamic (PD) endpoints**. These markers may or may not be related to the compound mechanism of action; however, they are susceptible to changes in drug exposure. These markers are typically applied in late preclinical and phase I human studies to select the appropriate dose of the therapeutic. Another class of biomarkers has been used to demonstrate **therapeutic efficacy**. In a 2004 study of drug attrition in human studies, the authors found that one of the most common reasons that compounds failed was due to lack of efficacy [3]. Early signs of efficacy can be determined using biomarkers related to the disease process. These biomarkers are extremely useful for chronic diseases that take years to present clinical signs of disease. An example in diabetes research is hemoglobin A1C. Reductions in HbA1C compared to placebo or standard of care provides the rationale for an approved intervention. The final group of biomarkers that have recently been applied to the drug development paradigm is **diagnostics**. These assays aid in the identification of patients with a specific disease. Diagnostic assays have their own approval process defined by regulatory bodies like the FDA. Laboratories that utilize these tools are governed and monitored by multiple governing bodies including the College of American Pathologists (CAP) and Clinical Laboratory Improvement Amendment (CLIA). Each of these types of markers has a different purpose in the drug development paradigm, they also has different regulations, and as a result may require different approaches to defining validity.

Selection of biomarkers for drug development depends on the objective of the measurement. Two terms that are routinely used to describe biomarkers are **proximal** and **distal**. A proximal biomarker is a chemical or biochemical measurement close to the molecular target of the drug and can also be referred to as a target engagement biomaker. A distal biomarker is a measurement that is downstream from the target and relates to the disease process. ELISA and flow cytometric assays specific for the phosphorylated residue at tyrosine 1068 have been developed. These assays would be considered a proximal biomarker for EGFR inhibitors. One of the effects of blocking the phosphorylation of EGFR in tumor cells is apoptosis. A distal biomarker of compound efficacy for EGFR inhibitors is cleaved caspase-3. The combination of these markers facilitates a better understanding of the therapy, because it is possible to demonstrate both direct inhibition of the target and the downstream biological effect, apoptosis.

The Biomarker Assay Development Process

The biomarker assay development process requires three very distinct and equally important components: biomarker selection, biomarker assay validation, and biomarker implementation.

The first step in the biomarker assay development process is selection. In order to determine the correct biomarker, a series of questions must be answered that relate to distance from the target, knowledge about the analyte of interest, and operational factors (Fig. 1). If the intent is to develop a pharmacodynamic biomarker, then a proximal or target engagement biomarker will probably be a better choice than a distal marker. An understanding of the biology is very important to design the best biomarker. Previous experience with the biomarker provides higher probability of technical success. Finally, operational factors must be considered such as analytical platform, reagent availability, and collection at the proper

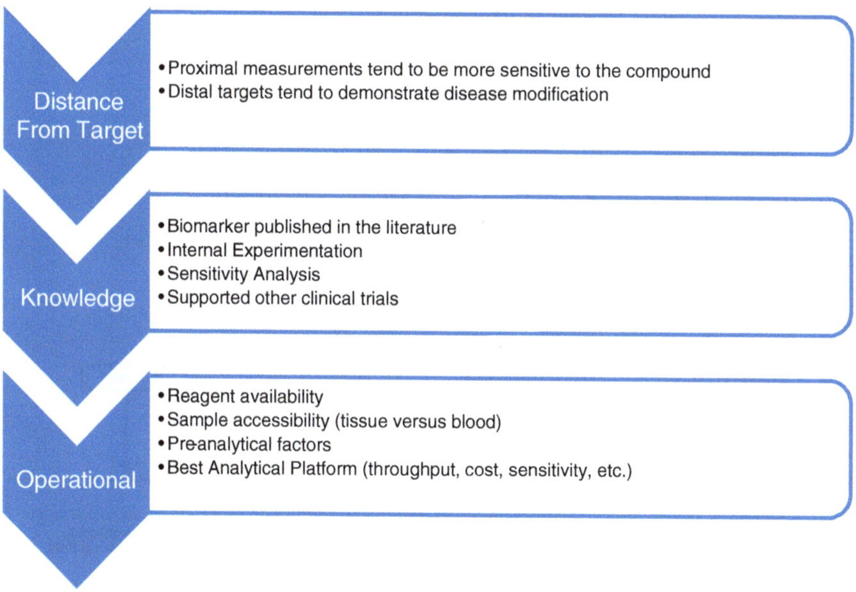

Fig. 1 Flow diagram of biomarker selection requirements

anatomical site. Once these factors have been addressed, then appropriate decisions can be made to enhance the likelihood of developing a biomarker assay that can meet the intended purpose. Figure 2 is a schematic representation of a decision tree that was developed in my laboratory to assist in the analytical platform selection for phosphoprotein assays to demonstrate kinase inhibition. In this setting, antibody availability, quantitative needs, and special factors such as single cell analysis or multi-analyte detection help to guide the selection process.

Once the biomarker has been selected, proper analytical validation is needed to determine if the marker meets the intended purpose. Analytical validation is a formal process of evaluating an assay method to ensure that it provides reliable data and is suitable for the intended application. Figure 3 represents a two parameter profile of risk versus analytical reliability. The factors and assumptions that influence the amount of analytical rigor are based on the type of decision being made, the stage of development, and the relationship between the marker and the disease. The quick test approach requires very little analytical validation and is used to identify large changes in a new or unproven biomarker. These methods are often applied in a shotgun screening approach to find new biomarkers. The second classification is the exploratory biomarker group. These are markers that are unproven clinically and have never been reported in a regulatory dossier, but may provide rationale for internal decision-making. Therefore, some level of reliability is needed to justify the decisions that are made. The last classification is the full

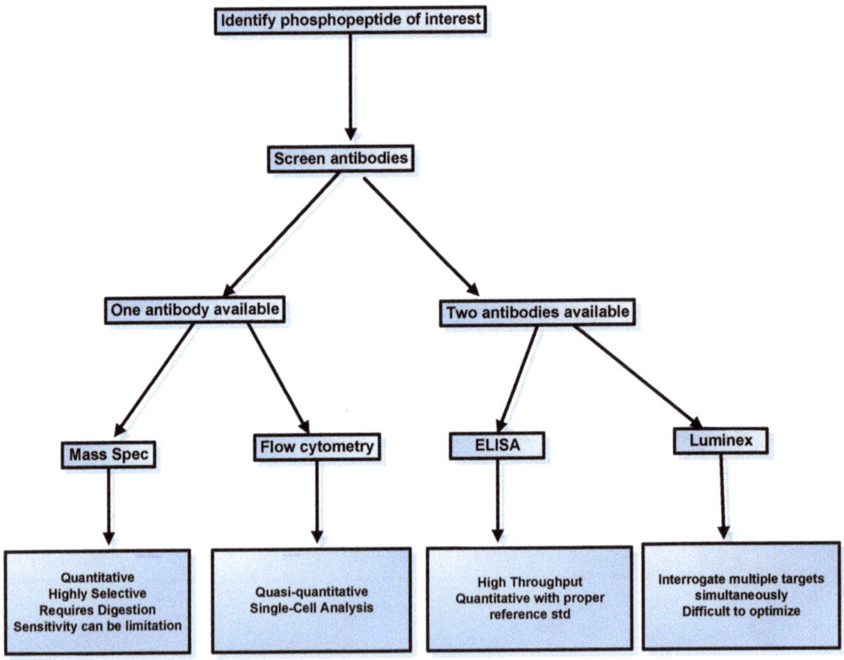

Fig. 2 Phosphoprotein biomarker platform selection

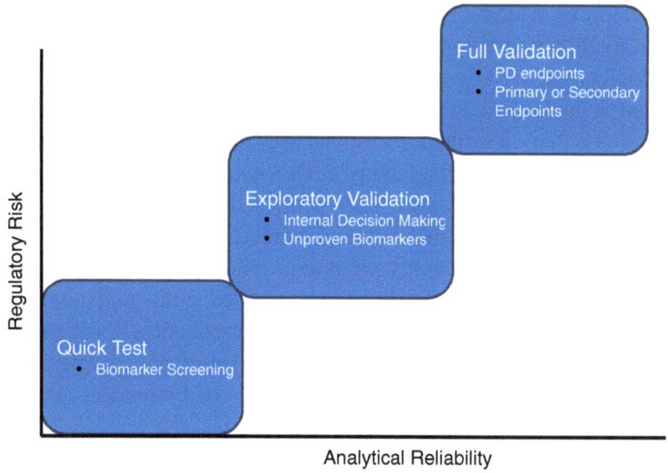

Fig. 3 Analytical validation risk assessment

validation. These are markers that will support regulatory filings as primary, secondary, or pharmacodynamic endpoints. These markers will provide the rationale for therapeutic efficacy and dose selection. As a result, these markers require a thorough understanding of the measurement reliability.

Biomarkers are endogenous, heterogeneous, and often structurally different from the calibrator [4]. It is for this reason that many in the field believed that additional nomenclature for classifying biomarker methods was needed. Lee et al., provided nomenclature for biomarker assays based on the availability of the reference standard and the type of response (continuous or categorical) [5]. Perhaps the single most distinct difference between biomarkers and drug measurements is the availability of a true reference standard. Within the continuous number category, there are three classes: definitive quantitative, relative quantitative, and qausi-quantitative. Definitive quantitative assays contain an absolute reference standard that is an exact copy of the analyte of interest. An example of a definitive quantitative assay is testosterone. Relative quantitative methods contain a calibration curve, but the calibrator is not an exact copy of the target analyte. One example of a relative quantitative method would be cytokine immunoassay using a recombinant protein standard that was produced in a bacterial expression system. The protein may lack the post-translational modifications and other features of the endogenous protein. Quasi-quantitative methods have a continuous response output; however, they do not have a calibration curve for determining concentration values. Flow cytometry and antibody titer determination are examples of quasi-quantitative methods. Categorical assays represent the last class of assays. These methods produce outputs of positive or negative. A genotyping assay to identify a point mutation is a good example of a categorical assay. Each class of biomarker has different requirements for validation.

When developing validation strategies for biomarker assays, we often look to the bioanalytical method validation guidelines and best practices for drug assays [6, 7]. The technologies that we use for drug assays are also applied to biomarker measurements, so it is logical to apply similar principles. Lee et al. [5] published recommendations that described the consensus approach for validating protein biomarkers using ligand binding assays. The authors advocated the concept of "fit-for-purpose" method validation. They proposed that the scientific rationale and project needs should guide the validation requirements. They also proposed learn-and-confirm model of in-study validation guiding on-going development and continued validation. In 2013, the US FDA released draft guidance that included sections on research use only biomarkers and the application of diagnostic tests for off label biomarker measurement [8]. The remainder of this chapter will focus on analytical validation strategies to address reliability of biomarker methods.

Exploratory Validation

In the assay development and exploratory validation setting, an understanding of what is being measured, how much error is associated with that measurement, and were the collection procedures properly conducted to avoid pre-analytical artifacts needs to occur. Biomarker scientists have a difficult challenge, because the analyte of interest is an endogenous molecule that may be present in normal individuals at measurable concentrations. The problem is discerning a true positive response in a test sample from a false positive signal caused by the sample matrix. On the other hand, biomarkers may also produce false negative responses caused by interfering substances in the disease sample. In both of these situations, how do you determine "truth"? In the field of biomarkers, the ability to demonstrate accuracy is more challenging than in other analytical disciplines. As mentioned previously, these analytes are endogenous, heterogeneous, and often structurally different from the calibrator. It is for these reasons that some in the field question the validity of traditional bioanalytical accuracy testing for biomarkers. I also believe that accuracy is more than just trueness, but represents the true in vivo concentration at the time of collection. Given the complications associated with measuring biomarkers, I recommend additional experiments that can complement a traditional spiked recovery experiment for assessing accuracy.

The ICH guidance states, *"The accuracy of an analytical procedure expresses the closeness of agreement between the value which is accepted either as a conventional true value or an accepted reference value and the value found.* ***This is sometimes termed trueness.****"* Two fundamental assumptions in bioanalysis are that you understand what is being measured, and the calibrator represents the target analyte. However, most protein biomarker assays use recombinant proteins from engineered systems, so the calibrator may not adequately represent the endogenous protein. Experiments that ensure the test system is capable of measuring the endogenous protein are essential. Ex vivo stimulation assays that produce cytokines, phosphoproteins, or growth factors at measurable concentrations can provide evidence that the method detects biologically relevant proteins [9]. For additional confidence, target modulation by inhibitors, agonist titration, or variable time of exposure improves confidence of a true response [10]. The combination of these two experiments provides confidence that the method is capable of detecting the protein of interest. Another important experiment that provides confidence in accuracy is orthogonal method comparison. If a gold standard assay is available, then the samples can be split and tested in both systems and compared for agreement [11]. Another elegant experiment that provides definitive characterization of what is being measured is an affinity capture, protein separation, and identification by mass spectrometry [12]. These experiments are often difficult to obtain given the limited resources, organizational constructs, technical expertise and availability of the platforms. However, this experiment provides the tools to describe the molecular species that were captured by the immunoassay reagent. These experiments provide greater understanding of the test method and what the assay is measuring.

Another important consideration for overall accuracy is sample collection, processing, and storage. Any of these steps can introduce artifact into the final result. There are several choices of additives and anticoagulants that prevent proteolytic activity of the clotting factors, prevent platelet activation, and provide a carbon source for whole blood samples. Depending on the anticoagulant selection, results can vary significantly [13]. Processing differences can also alter the final concentration. Platelets serve as reservoirs for many proteins found in circulation, so if the processing steps fail to account for the platelet contribution, then the results could be altered [14]. As a result, we recommend evaluating the effects of platelet rich and platelet free plasma to determine the best collection strategy. Storage is the final factor that can influence the result. One observation that I have made is that storage recommendations are assay specific for testing systems that use antibodies to measure a protein. The effects of storage can change the structure of the target protein and as a result change the antigen antibody interaction leading to differences in immunoreactivity. Therefore, it is essential to characterize the storage stability using the test method and not apply literature values, or other methods for the same protein.

Full Validation

The following parameters should be assessed during a full validation: Calibration/Standard Curve, Accuracy and Precision (Reproducibility), Limits of Quantification, Parallelism, Dilution linearity, Specificity and Interference, Stability, and Normal Range.

Calibration/Standard Curve

Biomarker assays for quantification require a standard curve. The concentration response relationship for immunoassays is a nonlinear response that is best fit with a nonlinear regression equation using either a four or five parameter logistic function. The relationship between response and concentration should be demonstrated to be continuous and reproducible. All assays must contain a minimum of 6 levels of nonzero standards in the valid assay range. Anchor points (calibrators outside the range of quantitation) should be added to improve the fit of the calibration curve. Selection of optimal calibration curve fit parameters will be determined during validation. Typically, a logistic regression model benefits from weighting, so careful examination of the best weighting factor must be conducted. Evaluation of the coefficient of determination, and the imprecision and bias of back-calculated standards, should be used to identify the model that provides the most consistent, accurate and precise data. The one algorithm with the best fit, and therefore most suitable, will be used throughout the validation and sample analysis. The standards are back-calculated and the mean back-calculated values should not deviate more than ±20.0 % from the nominal value at all concentrations (±25.0 % at LLOQ and ULOQ). The standards not meeting these criteria shall be excluded from the calibration curve. To accept a calibration curve, at least three-quarters of all standards analyzed in the run shall meet the stated acceptance criteria for a curve to pass acceptance criteria. Selecting the appropriate calibration matrix is a subject of great interest or debate. The options include buffer standard curves, modified or surrogate matrix, or authentic matrix pools. The buffer standard curve is simple to produce, easy to control, and is not flawed by potential endogenous biomarker contamination; however, careful examination of the minimum required dilution is needed to avoid artifacts in measurement of the test sample. The sample matrix could cause a reduction in immune-reactivity or an over quantification due to cross-reactivity of similar proteins or matrix components. The next option is modified or surrogate standard curve matrix. Examples of modifications are heat treating and charcoal stripping, affinity stripping, or alternative species (horse serum). This approach potentially provides a more consistent matrix composition than buffer; however, the process may remove important endogenous molecules that can influence the measurement, or introduce components that potentially cross-react with the target. These approaches can be laborious, expensive, and challenging to create consistency with different lots of calibrator matrix. The final approach of using authentic patient samples to prepare calibrator matrix pools requires undetectable biomarker levels, so this may not be feasible for many biomarkers that are present under normal physiological conditions. Careful examination of the specific biomarker will dictate the best approach and within a lab all of these approaches may be required.

Samples for Validation (and Quality Controls)

Another important consideration for biomarker assays is the selection of proper validation samples. Validation samples should be representative (in terms of matrix and species) of the samples to be assayed and cover the appropriate analytical range of expected results in the population to be studied. For rare species/matrices, attempts should be made to obtain representative material in at least one sample. There are two approaches to value assignment. The first option is referred to as methods of addition. The approach requires spiking recombinant analyte into a representative matrix and assigning a nominal concentration. It will be necessary at low concentrations to correct for the endogenous concentration. An alternative approach has been used in some laboratories and is based on the clinical diagnostic strategy for QC value assignment. The nominal concentration is determined empirically in authentic patient samples and applied throughout sample analysis.

Precision and Accuracy

Every analytical measurement has error associated with the value. Precision and accuracy analysis provides a metric that can be used to determine if the analytical error is acceptable to meet the goals of the study. In order to determine the analytical error, different components of the overall variation should be determined. The components of variation can be divided into parts that include inter-assay and intra-assay. The intra-assay component is the smallest discernible source of error that can be determined without isolating components of the assay process. This is accomplished by measuring multiple QC samples ($n = 3$) in a single analytical run. The inter-assay precision is composed of multiple parts that include operator, day, reagent lots, and batch. In order to adequately test the inter-assay precision a total of six individual batches over multiple days should be tested. Best practices recommend assaying quality control (QC) samples at five concentration levels spanning the entire range of the standard curve. Method acceptance should be based on the needs of the study design and the requirements of the biomarker; however, the majority of biomarker assays can tolerate 20–30 % CV and still provide acceptable data to inform decisions.

Accuracy assessments in the setting of full validation rely on spiked recovery experiments. For absolute or relative quantitative methods, QC samples are spiked at nominal concentrations and the measured concentration is compared to the nominal. A percent recovery or bias is determined. Unlike drug measurements, the endogenous biomarker component must be subtracted. There are two options for subtraction: (1) measure the blank pool over the course of the six analytical runs and subtract the mean; (2) subtract the blank value specific to each batch. Both of these options has a set of advantages and disadvantages; however, both options are introducing error into the final measurement which be augmented at the limit of quantification.

Limit of Quantification

The lower limit of quantification (LLOQ) is the lowest analyte concentration that can be quantified with acceptable precision and accuracy. The upper limit of quantification (ULOQ) is the highest analyte concentration evaluated that can be quantified with acceptable precision and accuracy. The LLOQ and ULOQ are expressed in concentration units of the reference material used to generate the calibration curve.

To determine the limits of quantification, QC samples prepared at the suspected LLOQ or ULOQ are analyzed along with the low, medium, and high QC samples ($n \geq 3$ at each concentration) in a minimum of six separate runs. Acceptable accuracy and precision will be dictated by the intended purpose, but the guideline has been 25–30 % total error at the LLOQ.

Another approach that is accepted in the clinical laboratory setting is to identify a truly blank patient sample and measure that sample a minimum of $n = 20$ results. The mean response and standard deviation should be determined from the 20 results. The limit of detection is defined as the concentration that corresponds to the response +3 standard deviations.

Biomarker Range Assessment

An important consideration for biomarkers is to understand the relative concentrations of the biomarker in the population of interest and in normal subjects. In order to evaluate the concentration range of the biomarker in the appropriate population; at least 30 individual patient samples should be analyzed. A distribution analysis should be conducted. The mean and various quartiles can provide an understanding of the distribution of concentrations. Typically, this approach is used for diagnostic assays; however, it can be useful for traditional biomarkers of efficacy, safety, and pharmacodynamics. The range assessment can assist study design and troubleshooting of aberrant in-study results. The range of results will provide insight into the biological variation that exists in the population of interest which will facilitate proper sample size calculation. The quartile analysis can also identify potential analytical or biological outliers that require additional investigation.

Parallelism

Parallelism is a condition in which dilution of test samples does not result in biased measurements of analyte concentration. Thus, when a test sample is serially diluted to result in a set of samples having analyte concentrations that fall within the quantitative range of the assay, there is no apparent trend towards increasing or decreasing estimates of analyte concentrations over the range of dilutions [4]. For biomarkers, parallelism tests provide assurance that the calibrator and the endogenous biomarker are sufficiently similar.

To evaluate the effect of individual sample matrix on the quantification of endogenous biomarker, a minimum of 6 individual samples should be tested for parallelism. Each individual sample will be tested with at least four dilutions.

Of the individual matrix samples that have at least 3 concentrations within the assay range, parallelism will be demonstrated by the recovery of dilution-adjusted concentrations relative to nominal endogenous level (as measured in the undiluted condition) should be within ±20.0 % (±25.0 % for samples measured between LLOQ and LQC levels at a given dilution), with % CV between the back-calculated concentrations of duplicates $\leq$20.0 %. There are other statistical tests that can be applied to parallelism assessment including an ANOVA or F test; however, each of these requires tools that are generally not included in immunoassay regression software. The other problem with these approaches is a lack of consensus among statisticians and experts on the best method for testing parallelism. In the cases where parallelism fails, decisions must be made whether the assay can be used at a fixed dilution factor or redeveloped with different antibodies or a different calibrator.

Dilutional Linearity

Another important validation parameter is dilutional linearity. Some biomarkers are elevated in diseased patients and require sample dilution to achieve measurable concentrations. Dilutional linearity experiments demonstrate that spiked samples can be diluted accurately across a range of dilutions that correspond to different areas of the standard curve. The dilutional linearity experiment does not replace the requirement to conduct parallelism experiments when suitable sample matrix becomes available. Parallelism assessment may require the use of incurred samples. The interpretation of dilutional linearity is similar to that of parallelism; however, there is a nominal concentration after dilution adjustment. Figure 3 is a schematic representation of linearity and nonlinearity. The same dilution-adjusted mean concentration and %CV can be computed for linearity.

Biomarkers in Drug Discovery and Development: Pre-analytical and Analytical Considerations

R. Aleks Davis, Andrew P. Mayer and Ronald R. Bowsher

Abstract Today biomarkers are used widely across the drug development continuum to improve the success rate and cost-effectiveness of rational drug development [1, 2]. These include diagnostic indicators of disease, assessments for biological pathways for new therapeutics, confirmation of target engagement or mechanism-of-action, safety indicators, and as pharmacodynamic (PD) measures of therapeutic efficacy [1, 2]. The application of biomarkers is now commonplace in drug development and has helped usher in a new era of predictive, preventative, and personalized medicine, where therapeutics can be tailored to a patient's unique biology. However, prior to being used for these purposes, there are several practical aspects of biomarker utilization that need to be taken into account. These include biomarker validation and qualification. Analytical validation is the formal process of evaluating an assay to ensure it provides reliable data [1, 2], whereas qualification (also known as clinical validation) is the evidentiary process of linking a biomarker with biology, pharmacology, or clinical endpoint [1, 2]. Analytical validation is important, as issues of quality herein can limit the utility of biomarker data. Moreover, a lack of proper analytical validation can potentially undermine successful demonstration of a biomarker-related pharmacodynamic effect during clinical investigation [1, 2]. Not surprisingly, successful biomarker testing in drug development requires careful consideration of pre-analytical factors that may impact reliable biomarker quantification in biological matrices. Most often, pre-analytical factors include activities associated involving collection, shipment,

R.A. Davis (✉)
B2S Life Sciences, Franklin, IN, USA
e-mail: aleks.davis@b2slifesciences.com

A.P. Mayer
GlaxoSmithKline Pharmaceuticals Research and Development,
King of Prussia, PA, USA
e-mail: andrew.p.mayer@gsk.com

R.R. Bowsher
B2S Consulting and AIT Bioscience, Franklin, IN, USA
e-mail: ron.bowsher@b2slifesciences.com

R. Weiner and M. Kelley (eds.), *Translating Molecular Biomarkers into Clinical Assays*, AAPS Advances in the Pharmaceutical Sciences Series 21,
DOI 10.1007/978-3-319-40793-7_2

tracking, storage, and distribution of test samples. Analytical considerations include the use of kits versus custom assays, reference standard sourcing, sample preparation strategies [3, 4], instrumentation and the removal of matrix interferences. This chapter will discuss some of the various factors that need to be taken into account when preparing for biomarker analysis.

Keywords Pharmacodynamics (PD) measures · Safety indicators · Validation · Sample collection kits · Storage · Shipping

Biomarkers in Discovery and Development

Biomarkers are useful tools throughout the development cycle of a therapeutic; however, the rigor to which the biomarker assay is tested will change depending on the stage of development as well as the intended use of the data. This idea of a "fit for purpose" validation of biomarker assays is detailed in Lee et al. [5] which defines the validation of a biomarker assay as an iterative process, where the extent of validation increases as the therapeutic moves down the pipeline. Validation is broken into two phases, exploratory method validation and advanced method validation. The exploratory phase is less rigorous and would be used for biomarker in the preclinical and early clinical stage. An advanced validation would be undertaken when a biomarker is in late phase clinical, especially when the biomarker is used as a primary endpoint of a clinical trial. Biomarker assays used in discovery would often have very little characterization, however, would give an indication of up and down regulation of various biomarkers.

Pre-analytical Considerations

The ability to accurately detect and report biomarker results relies on executing several pre-analytical activities in a manner that limits their inherent variability. From the collection of the sample at the clinical site, all the way through shipping, storage and analysis there are several operational and logistical steps that require particular attention so that the sample conditions are consistent and traceable. Each point along the sample supply chain, no matter how insignificant it may seem, presents potential opportunity for inconsistent handling or mismanagement. In order to mitigate the risk of variability, the complete supply chain must be validated to ensure the appropriate steps are in place for every point along the sample's lifecycle.

Collection

The biological sample will almost always be collected at a clinical site. At this phase it is necessary to ensure the staff have proper training in the desired method of collection, labeling, and outbound shipping.

Sample collection kits can be supplied to the site staff which ensures the proper equipment and supplies are used for specific sample collection needs [1]. Kits should include specific collection tubes for the type of matrix that is needed. They can also include needles, protective supplies for the phlebotomist, Band-Aids and gauze, and sample shipping materials.

Proper labeling of the collection tubes are going to ensure the specimen is tied back to the patient and not inadvertently switched after collection. Labels come in a variety of sizes and styles. Typical components include an accession ID, some iteration of subject and study identifier, the visit ID and a barcode so that registration can be streamlined. Tubes can come prelabeled or unattached within the kit and each comes with its own advantages.

Also included in each kit, should be a manual or set of instructions. The manual should clearly identify the visit and collection schedule, a list of components in the kit, instructions on how to use the components, and how to process, store and ship the samples. The manual should be very clear about when samples are collected throughout the day. Since many biomarkers are found to be dynamic, varying in concentration throughout the day [2], and can change with dietary influence, this need to be covered specifically in the manual. Variation in collection times can lead to variation in analytical results.

Biomarker collections come from all over the world. Providing the sites with a consistent set of supplies and instructions helps to lessen the variability and ensure a high level of confidence in the results.

Shipping

The method of shipping needs to be properly qualified prior to the samples being collected. Different shipping containers are rated to allow for specific temperature variations. The location of the site and shipping route can greatly affect the materials needed for shipping. Whether the samples are required to be shipped ambient, refrigerated, or on liquid nitrogen; it is important to understand the tolerable temperature limits of the sample. With those limits in mind, the selection of shipping container will need to withstand both high and low fluctuations [6].

Ensuring the shipping conditions are within the limits of the sample can happen prospectively and retrospectively. A prospective qualification of the shipping means data is collected prior to the shipment of the samples to better understand the environmental factors that will influence the container. You must map the fluctuation of temperature throughout the day and night and since conditions change with the seasons, yearly temperature readings need to be included in the assessment. Since the supply chain of samples is global, measurements must be made along the whole shipping route. Alternatively, data loggers can be included in the ongoing shipments, providing complete traceability of the sample throughout the trip.

Shipping biological samples internationally comes with the risk of encountering customs delays due to transportation regulations. When the shipping containers are selected, ensure there is adequate room on the surface to display the appropriate warnings and labels necessary. A container that is held up at customs due to incorrect labeling will not necessarily be in a position to be re-filled with ice pack or dry ice. For this reason, a shipment notification system should be in place to effectively communicate the departure and expected arrival of these packages.

Storage

The sample will likely spend most of its time sitting in storage. For that reason, it is important for the environment or unit to be properly temperature controlled and the location of the sample be clearly defined. Whether the samples are stored ambient or at −80 °C, the controlled areas will need to be monitored 24 h a day. Additionally, the monitoring system needs to have temperature thresholds that are connected to audible and traceable alarms notifications that record any temperature deviation.

The recording of temperature deviations in a system gives the ability pin down exact moments of temperature variation in the storage life of the sample. Know biomarker stability data can be cross-referenced against the temperature log, in case of questionable results. For this reason, the IT system that is used needs to be username and password protected, ensuring an accurate audit trail of the sample. The system should also be connected to a calibrated temperature probe that is separate from the unit's original probe.

Even within enclosed temperature controlled environments, there can be significant variation throughout the space. Temperature mapping takes into account the temperature fluctuation of the area. Calibrated temperature data loggers should be used to record temperature in predetermined points with the 3D space over the course of 24 continuous hours. Data should be collected at different points along the height, length and depth of the unit and the total amount of probes is dependent on the total volume of the area [7]. Finally, the data should be compared to the set point of the unit, and the monitoring probe should be placed in the location of the unit where there is the highest level of variance. Placement of the probe in this location ensures the samples will be within the allowable limits of the temperature settings. The location of highest fluctuation will likely be different in every unit, thus the monitoring probe will be in a different position as well. If the probe is not placed in this precise location, the result will be samples potential exposed to temperatures outside of the allowable limits without any indication from the temperature alarm system.

Sample Tracking

The samples need to be tied to specific data that can be tracked along the lifecycle of the sample. The system that is being used to track the samples will also need to capture associated data such as aliquot quantity, volume, and concentration (if necessary). If there are known analyte stability issues as a result of freezing and thawing, the sample may be aliquoted prior to being placed into storage. Having a system in place that accommodates the need of connecting sample lineage, will make organization of the samples easier. If aliquots are necessary, any data associated with the parent sample will need to be related to the aliquot as well.

Any changes in the data should be racked through a verifiable audit trail. Changes in subject information, label information, and sample data can lead to confusion at the time of analysis and results can be misassociated with the wrong subject [8]. Sample locations should also be tracked down to the most specific locations they can be. Pulling many samples from storage at once can be tedious and error prone, particularly if the exact locations are not known. Storage units should be situated with unique positions that allow for single samples to be placed and easily identified when they are later pulled. Providing specific locations allows for quicker access in and out of the freezer, reducing the amount of time the doors are ajar exposed to outside temperatures.

Any movement of the sample should be demonstrated in a chain of custody report. The chain of custody report needs to have detailed information such as personnel ID's of individuals working with the samples, confirmation of sample receipt, storage locations, shipment locations, and time in and out of possession. This data confirms the correct storage and handling of the sample at each step along the lifecycle.

Distribution

Whether the analysis is being performed in close proximity to the storage location or the samples are being shipped, a detailed manifest should accompany the package. The manifest should include position within the package, and an identifier that corresponds to the tube label. If samples are inadvertently placed in the wrong location, the manifest data and tube can be reconciled against one another and sorted out.

Sometimes, samples are stored in a 96-well plate format and given the emergence of automated analytical technologies, the samples can often be delivered in the same format. There are some 96-well plates that hold individual tubes that can be placed in varying orientations. These tubes typically have enough surface area to support a labeling system or a 2D barcode on the bottom of the tube that can be tied to the manifest. In the cases where individual labels are not available for each well, a rigorous QC step need to be employed so that the sample location are accurate and traceable from the manifest.

Analytical Considerations

Several analytical characteristics need to be examined when utilizing a biomarker to support drug discovery. One question to investigate prior to beginning lab work is the use of a commercial kit versus a custom built assay. Commercial kits are attractive due to the perceived ability to use them "off the shelf"; however, one must test these kits to determine if they are truly fit for their intended purpose. Often times the kit must be modified to gain acceptable performance, however, this can be difficult when critical information is not known, such as concentrations of antibody pairs and the components in the kit assay buffer. When sourcing commercial kits and reagents, enough material should be purchased to limit the number of lot-to-lot changes during the course of a study. The manufacturer of the material is also important, as some vendors are more reputable than others. While the development time needed for a custom assay is typically more, there is greater flexibility with the custom assay. The antibody pairs can be optimized for the desired quantitative range of the assay and an assay buffer can be developed which helps to reduce interferences in the matrix of interest. Another consideration prior to starting lab work is the endogenous concentration of the biomarker. While some biomarkers are high in concentration, most are relatively low and require a sensitive assay for measurement. This is critical when selecting an instrument platform to be used. While a standard ELISA with absorbance detection is simple and cost effective, it may be impossible to obtain the desired sensitivity. One solution is to change out the enzyme substrate for a more sensitive detection scheme. When using horseradish peroxidase as the enzyme detection system, one can switch substrates from TMB (3,3′,5,5′-Tetramethylbenzidine) to ADHP (10-Acetyl-3,7-dihydroxyphenoxazine) for fluorescence detection or luminol for chemiluminescence detection. The same process can be applied to other enzymes as well. If even more sensitivity is needed, one can use the Meso Scale Discovery platform with electrochemiluminescence detection.

The source of the biomarker reference standard is also an important point to consider. Unlike small molecules, which are synthesized and have a defined purity, biomarkers are endogenous with various post-translational modifications and the extent of glycosylation can differ from patient to patient. Thus, it is impossible to have a reference standard which is indicative of the analyte be measured in vivo. Testing spike-recovery and assessing parallelism can be of great importance, helping to determine the extent of homogeneity between the endogenous biomarker and reference material. The desired matrix must also be explored, as serum, plasma, and different anticoagulants can have an effect on your assay. If stability is a concern, one should steer away from serum, as the length of time needed to clot can cause stability concerns, and it is not amenable to protease inhibitors. Na/Li heparin plasma has been shown to cause interferences in ELISA assays and is rarely used. Potassium EDTA (both K_2 and K_3) are commonly used, however, keep in mind that K_3 EDTA plasma has been shown to effect pH and both can have a negative effect when alkaline phosphatase is used as the enzyme system. In addition, additives to the matrix need to be assessed, especially for smaller proteins and peptides. Some peptides can be especially prone to proteolytic degradation, and thus the addition of a protease inhibitor cocktail is crucial for reproducible results.

A major complication when setting up a biomarker assay is the effort to remove matrix interferences [9, 10]. There are many things to examine, including the species of the antibodies, degree of post-translational modifications (sugars), and the presence of heterophilic antibodies and rheumatoid factor. The addition of serum components to the assay buffer can aid in the removal of nonspecific interactions. A good first step is to identify the species of the antibodies used in the assay, keeping in mind that the total serum content in the assay buffer should be roughly 5 % [10, 11]. If a noncompetitive ELISA is being used and the capture antibody is raised in mouse and the detection antibody is raised in goat, then the assay buffer might contain 5 % mouse serum and 5 % goat serum. If both the capture and detection antibodies were raised in mouse, then the assay buffer would instead generally contain 5 % mouse serum. Addition of other serum sources can be used as well; we have had good results with the addition of equine serum. Knowing the degree of post-translational modification of your antibodies and types of sugars can give a clue as to additional buffer components. Sugars bound to your antibodies can often be very sticky and the addition of sugar to your assay buffer can help to reduce these interactions [12, 13]. Heterophilic antibodies often cause problems in immunoassays. Heterophilic antibodies are typically low affinity binders and are polyreactive against poorly defined antigens of different chemical composition [9, 10, 14–17]. While they may be low affinity binders, they are present in 20–40 % of subjects [10]. Known generally as Human Anti-Animal Antibodies (HAAA), these anti-species antibodies create both false-positive and false-negative results [13]. This is an area where the addition of serum from the species of your detection antibodies can greatly aid in the reduction of this phenomenon, as the anti-species heterophilic antibodies will interact with the free serum components as opposed to the antibodies. Knowing the typical patient population's disease state can also aid in the choice of assay buffer components. Rheumatoid factor, for example, is often found in rheumatoid arthritis patients, and can cause false-positive results due to bridging of the capture and detection antibody pairs.

An analytical consideration often overlooked by immunoassay scientists is the use of an extraction procedure for sample cleanup prior to the immunoassay. An extraction step is not practical for larger proteins, since extractions are intended to remove proteins, which are often the analyte of interest. However, for peptides, extractions can be useful in removing endogenous interferences from the immunoassay. This can be especially critical for immunoassays of peptides, since specific antibodies against peptides can often be difficult to generate. In addition, there are often matrix binding proteins which interfere with measuring the true endogenous component. Low affinity binders can include albumin and circulating free protease inhibitors. High affinity binders can include soluble receptors and anti-drug antibodies (ADA). There are several extraction reagents which can be used. Extractions with polyethylene glycol (PEG) are an attractive choice, since the extraction reagent is in the aqueous phase, and the supernatant can be directly applied to the immunoassay. However, sometimes PEG can cause interactions with the antibodies in the immunoassay, reducing a signal, and rendering them useless. Another option is the use of alcohols and acetonitrile. These reagents can be used to crash out the proteins. The resulting supernatant can be dried and reconstituted in the assay buffer of choice. Yet another option is the use of an acid treatment step. The addition of an acid to the samples, reducing the pH to ~2–3, can help to break apart the interactions with binding proteins and ADA. However, care must be taken to ensure the pH of the resulting reconstituted extract is neutral, as the addition of an acidic supernatant to the immunoassay can reduce the biding of analyte with the antibodies. This can be done by increasing the ionic strength of assay buffer or by incorporating a neutralization step prior to the drying down the sample.

As discussed previously, the extent of validation is dependent upon the intended use of the biomarker data. Regardless, there are several experiments that should be performed to ensure the assay is performing adequately. One question that needs to be answered almost immediately is the endogenous concentration of the biomarker in vivo. Often times, a surrogate matrix will be used in lieu of a matrix calibration curve. For biomarkers with high endogenous concentrations, it can be difficult to source matrix with low enough levels to be used for calibration. This surrogate matrix can be stripped to remove the endogenous analyte of can consist of a protein buffered solution. If a surrogate matrix is used, it is important that quality control samples in the matrix of interest are included during the validation to ensure that the biomarker is being recovered from the matrix. Depending upon the extent of validation, anywhere from three to six runs to determine precision and accuracy can be run to determine this important parameter. A selectivity (spike/recovery) experiment should be performed in various individual lots of matrix to ensure that spiked analyte is being recovered. Each lot should be analyzed with both spiked and unspiked with drug to account for the endogenous concentrations. Dilutional linearity should be performed to ensure that samples with high concentrations can be diluted into the range of the curve, and that no hook effect is observed. Parallelism, as discussed previously, is important to ensure that endogenous analyte, when diluted, is parallel to the calibration curve. When possible, the specificity of the critical should be defined, by spiking similar analytes into control samples and assessing recover of the biomarker of interest.

Conclusion

There are many areas to consider when preparing to run a biomarker assay. From the point of collection to the moment of analysis, there are multiple points of variability that can influence the outcome. Taking the time to consider these prior to execution will result in more consistent and reliable data.

References

1. Bowsher R, Nowatzke W (2012) Application of commercial research-grade biomarker assays in drug development: is it time to create 'pharmaceutical grade' kits? Bioanalysis 4(20): 2427–2430
2. Nowatzke W, Cole T, Bowsher R (2012) Systemic analytical validation of commercial kits for the determination of novel biomarkers for clinical drug development. Bioanalysis 2(2): 237–247
3. Tuck M et al (2013) Serum and plasma collection: preanalytical variables and standard operating procedures in biomarker research. Proteominc Metab Approaches Biomark Discov Chap 5:78–85
4. Ye X et al (2013) Tissue sample preparation for proteomic analysis. Proteominc Metab Approaches Biomark Discov Chap 3:40–49
5. Lee JW et al (2006) Fit-for-purpose method development and validation for successful biomarker measurement. Pharm Res 23(2):312–328
6. Holland N, Bastaki M et al (2005) Molecular epidemiology biomarkers—sample collection and processing considerations. Toxicol Appl Pharmacol 206:261–268
7. Tworoger S, Hankinson S et al (2006) Collection, processing, and storage of biological samples in epidemiologic studies: sex hormones, carotenoids, inflammatory markers, and proteomics as examples. Cancer Epidemiol Biomark Prev 15:1578–1581
8. Campbell L (2012) Best practices for repositories: collection, storage, retrieval and distribution of biological materials for research. Biopreservation Biobanking 10:81–161
9. National Metrology Laboratory (2012) User guide to mapping temperature controlled enclosures 4:1–8
10. Wagar EA, Stankovic AK, Raab S, Nakhleh RE, Walsh MK (2008) Specimen labeling errors: a Q-probes analysis of 147 clinical laboratories. Arch Pathol Lab Med 132(10):1617–1622
11. Tate J, Ward G (2004) Interferences in immunoassays. Clin Biochem Rev 25:105–120
12. Selby C (1999) Interference in immunoassay. Ann Clin Biochem 36:704–721
13. Kricka L (1999) Human anti-animal antibody interferences in immunological assays. Clin Chem 45:942–956
14. Hamilton RG, Adkinson NF (1985) Naturally occurring carbohydrate antibodies: interference in solid-phase immunoassays. J Immunol Methods 77:95–108
15. Adedoyin J, Johansson SGO, Gronlund H, van Hage M (2006) Interference in immunoassays by human IgM with specificity for the carbohydrate moiety of animal proteins. J Immunol Methods 310:117–125
16. Levinson SS, Miller JJ (2002) Towards a better understanding of heterophile (and the like) antibody interferences with modern immunoassays. Clin Chim Acta 325:1–15
17. Bouvet JP, Quan CP, Dighiero G (2002) Polyreactivity is not an artifact. J Immunol Methods 254:199–201

References

1. Bowsher R, Nowatzke W (2012) Application of commercial research-grade biomarker assays in drug development: is it time to create 'pharmaceutical-grade' kits? Bioanalysis 4(20): [illegible]

[illegible]

Biomarker Discovery

Omar F. Laterza and Xuemei Zhao

Abstract Biomarkers are used throughout the drug development process from target identification and validation to post marketing studies. Biomarkers are most impactful in informing dose selection, pathway associated activity and toxicity, clinical efficacy and in identifying subsets of patients who will most likely benefit of a therapy or who will develop a disease-related outcome. Great efforts have been placed in identifying novel biomarkers that may enable drug development. For this, a variety of technologies such as genomics, proteomics, metabolomics, cell-based flow cytometry, etc., are routinely used in the discovery of molecular biomarkers. In this chapter, we discuss the approaches used in discovery of different types of molecular markers and the utilities of these markers to facilitate drug development.

Keywords Biomarker discovery · Molecular biomarkers · Genomics · Proteomics · Metabolomics · Flow cytometry · Profiling

A biomarker is defined as a "characteristic that is objectively measured and evaluated as an indicator of normal biological processes, pathogenic processes, or pharmacological responses to a therapeutic intervention." This definition encompasses not only molecular biomarkers but also imaging biomarkers such as MRI, PET, and X-ray and other metrics such as blood pressure [1]. This chapter will focus on the different technologies used for the discovery of **molecular biomarkers**, which refer to "the subset of biomarkers that might be discovered using genomics or proteomics technologies." An expansion of this definition was proposed in 2007 to include those biomarkers "that are measurable or detectable based on their molecular characteristics" [2]. This wider definition include chemical entities such as nucleic acids, peptides, proteins, lipids, metabolites, and other small

O.F. Laterza (✉) · X. Zhao
Translational Molecular Biomarkers, Merck Research Laboratories, Rahway, NJ, USA
e-mail: omar_laterza@merck.com

X. Zhao
e-mail: xuemei_zhao@merck.com

R. Weiner and M. Kelley (eds.), *Translating Molecular Biomarkers into Clinical Assays*, AAPS Advances in the Pharmaceutical Sciences Series 21,
DOI 10.1007/978-3-319-40793-7_3

molecules that are measurable in body fluids and tissues (collectively termed "matrices") such as blood, plasma/serum, cerebrospinal fluid, bronchoalveolar lavage fluid, biopsies, etc. As in the above definition, a biomarker encompasses not just a measurement of a chemical entity obtained in a given matrix, but also a biologic context that defines its utility. Accordingly, "biomarker discovery" means the identification of chemical entities that are intimately associated to a biological/physiological process. The novelty implied by the term "discovery" may not necessarily refer to a chemical entity per se, but rather to the intimate association between the chemical entity and a biological/physiological process, which is defined above as a biomarker, and this association will be used in disease assessment as well as drug discovery and development.

Classification of Biomarkers

Biomarkers are normally classified based on their applications. In laboratory medicine, biomarkers are classified as: diagnostic, prognostic, disease staging, or for monitoring the clinical response to an intervention. In pharmaceutical development, specific response biomarkers inform on safety and pharmacodynamic (PD) responses. PD biomarkers are indicators of either a proximal (e.g., target engagement or immediately downstream) or distal physiological response, and are frequently used to predict dose responses.

Recently, with the increased desire of precision medicine, certain biomarkers are being used to identify sub-populations that are more likely to benefit or suffer (an adverse event) from a therapy. These biomarkers, referred to as companion diagnostics, are co-discovered and co-developed with the pharmacological agent and are subject to regulatory approval. They are ultimately deployed in clinical laboratories throughout the world. One specific field in which they have been applied is targeted therapies for cancer. An example of this type of biomarker assay is the PathVysion HER-2 DNA Probe Kit (PathVysion Kit) by Abbot Molecular, Inc. This assay detects amplification of the HER-2/neu gene in formalin-fixed, paraffin-embedded human breast cancer tissue specimens using fluorescence in situ hybridization (FISH). The Pathvysion Kit is indicated as an aid in the assessment of patients for whom herceptin treatment is being considered. The number of companion diagnostic tests will continue to expand in the future. A comprehensive list of these tests could be found at the following website: (http://www.fda.gov/MedicalDevices/ProductsandMedicalProcedures/InVitroDiagnostics/ucm301431.htm).

Discovery of Molecular Biomarkers

In general, biomarker discovery follows one of two approaches. The first is large, unbiased, 'profiling' involving thousands of different genes, proteins, metabolites or 'features', analyzed in an agnostic fashion. While in the second, experiments focus on prespecified biomarkers based on some previous knowledge on the disease. For instance, when trying to identify biomarkers for an inflammatory disease, the search may focus on a panel of inflammatory mediators (cytokines); or when working on an oncologic drug, the search may focus on genes/proteins associated with cell growth, proliferation, angiogenesis, etc. In either approach, at least two different populations are compared to each other (i.e., disease versus healthy control). It is extremely important, when performing these experiments, that the populations are well characterized and the number of variables is minimized (age, gender, ethnicity) or taken into account when performing statistical analysis. Also, sample collection, processing, and storage should be well standardized and controlled to minimize pre-analytical variability.

Below, we will briefly describe different technologies used in biomarker discovery, providing examples of biomarkers that have had an impact in the clinic. The in-depth description of these technologies can be found in other chapters of this ebook.

Biomarker Discovery in Genomics: The completion of the human genome project, as well as the continuing emergence of genomics technologies has resulted in an explosion of discoveries in the field of genomics.

Genomics biomarkers can largely be classified as gene expression (mRNA) biomarkers and biomarkers based on DNA mutations, polymorphisms or other genetic variants. Often, a number of genomics biomarkers are predictive when used in combination. These are referred to as gene signatures and there is generally an algorithm applied to the data analysis in order to ultimately determine the diagnosis or prognosis of the patient or to predict the response to a therapeutic intervention. For example, a 70-gene signature was found to be strongly predictive of a poor prognosis in lymph-node negative breast cancer patients [3]. The signature was built using DNA microarray technology on samples from 117 young patients, and applying a supervised classification technique. The gene signature, which included genes regulating cell cycle, angiogenesis, metastasis and others, outperformed the existing clinical parameters in predicting disease outcome and enabled selection of patients who would benefit from adjuvant therapy. This signature was further validated in a retrospective clinical study in which 295 lymph-node negative patients were studied. Multivariable Cox regression analysis showed that the prognosis profile was a strong independent factor in predicting disease outcome [4].

Other genomics biomarkers include mutations and single nucleotide polymorphisms (SNP), which are typically, identified through genetic linkage analysis or genome-wide association studies (GWAS), which assess the association of individual SNPs with patient characteristics across a population. For instance, hypersensitivity to the HIV drug Abacavir was found to be linked to the HLA-B*5701 polymorphism by performing linkage analysis [5]. In another study, polymorphisms linked to response to anti-HCV treatment were identified. It had been known for some time that people of European decent have a significantly higher probability of being cured than patients of African descent when receiving the recommended treatment for chronic HCV infection, which includes a 48-week course of peginterferon-α-2b (PegIFN-α-2b) or -α-2a (PegIFN-α-2a) in combination with ribavirin (RBV). Through GWAS analysis, a genetic polymorphism near the *IL28B* gene, encoding interferon-λ-3 (IFN-λ-3), was found to be associated with an approximately twofold difference in response to treatment. The genotype predicting better response is present in substantially greater frequency in European populations.

Other examples include the mutations in the BRCA1 and BRCA2 genes. Mutations in the BRCA1 gene, and their association with increased risk in breast cancer, were first reported in 1990 as a result of DNA linkage studies. In 1994, a gene similar to BRCA1, named BRCA2, was discovered and mutations in this gene were also found to be associated with a higher risk for breast cancer [6].

Both of these genes encode tumor suppressor proteins which are involved in DNA repair mechanisms. Mutations in both genes are associated with increased risk for breast and ovarian cancers as well as for several other cancers. **Another type of genomics biomarkers that has been intensely investigated in the last few years are microRNAs (miRNAs)**. miRNAs are small non-coding RNA molecules that regulate mRNA translation by either causing degradation of specific mRNA or blockade of their translation. Some of them are expressed in a tissue- and disease-specific manner. These properties, in addition to the fact that they are rather stable in plasma and are amplifiable, make them very promising potential biomarkers in the diagnosis and prognosis of human diseases.

miR-122 has been investigated as a biomarker of liver injury. In animal models of drug-induced liver injury, miR-122 exhibited a >6000 fold increase suggesting that miR-122 may be a specific and sensitive indicator of acute hepatocyte damage [7]. In a small proof-of-principle study, plasma miR-122 levels were determined in 15 apparently healthy volunteers and in 30 patients with documented liver disease. miR-122 concentrations were statistically different in patients diagnosed with hepatitis C virus versus healthy volunteers (193.8 vs. 51.7 copies/20 pg RNA, respectively, $P = 0.0015$), as it was in patients with other liver diseases (202.3 copies/20 pg RNA, $P = 0.028$) [8].

miRNAs may be linked to the pathogenesis of diseases. For example, miR15 and miR16 were found to contribute to chronic lymphocytic leukemia (CLL). Since then other miRNAs have been reported in the context of other cancers, rheumatoid arthritis, ectopic pregnancy, and others [9].

From Genomics to Proteins and from Mouse to Man in Biomarker Discovery

In an interesting approach, **gene expression** profiling was also used as the basis for the discovery of brain-specific **proteins** that could be used in the diagnosis of stroke and neurodegenerative diseases [10]. The study acquired mRNA profiles from every major organ of a mouse using microarray technology to identify mRNAs differentially and abundantly expressed in the mouse brain. The purpose of this approach was to identify specific proteins released upon brain cell death into the extracellular space, the CSF, and ultimately in the circulation. These abundant and brain-specific proteins would be ideal biomarkers of brain cell death.

After the identification of the brain-enriched mouse **genes**, sensitive immunoassays were developed to measure the corresponding human **proteins**. One of these proteins, Visinin-Like Protein 1 (VILP-1), was detected in the CSF of rats that were subjected to middle cerebral artery occlusion, a rat model for stroke, whereas its levels were undetectable in normal rat CSF. Furthermore, VILIP-1 was also detected in the plasma of human patients that had suffered a stroke, but not in healthy control subjects. These experiments demonstrate how animal models and tissue-specific mRNA profiling may help in the discovery of protein biomarkers in the circulation. At that time, the discovery of VILIP-1in CSF as a marker of brain cell death by LC-MS-based proteomics profiling would have been extremely difficult or impossible given the limitations in the sensitivity of the technology.

Protein Biomarker Discovery by Profiling

Protein biomarkers have also been identified using an unbiased proteomics profiling approach. Proteomic approaches can systematically profile, identify, and quantify proteins expressed in a given cell or tissue and quantify and identify relative changes between two or more samples representing different conditions [11]. From a biomarker discovery perspective, proteomics provides the most utility when it compares protein expression profiles of samples in different biological states, such as disease versus normal or drug- versus placebo-treated, to identify proteins that are differentially expressed in one condition compared to the other. The initial approach was carried out by two-dimensional gel electrophoresis coupled with mass spectrometry (MS) analysis. More recently, quantitative proteomics—stable isotope labeling-based and label-free proteomics—has emerged as technologies with greater speed, throughput, robustness, and sensitivity.

Stable isotope labeling-based methods such as ICAT (isotope-coded affinity tag) [12], iTRAQ (isobaric tags for relative and absolute quantitation) [11], and SILAC (stable isotope labeling with amino acid in cell culture) [11] incorporate heavy versions of specific molecules into peptides or proteins either by chemical derivatization or by metabolic labeling. Every peptide appears as a pair with a defined mass difference from the heavy and light isotope-labeled conditions. The intensity ratio of a peptide pair measured by LC-MS represents the ratio of peptide/protein abundance in two biological conditions. Thus, the abundance of peptide/protein in different conditions can be compared. A powerful but more challenging approach for biomarker discovery is label-free differential MS (dMS). In addition to reproducible biochemical sample preparation to isolate relevant sub-proteomes and LC-MS profiling data acquisition, the dMS approach utilizes a sophisticated software tool (dMS) to compare intensities for each peptide feature (retention time and mass-to-charge ratio *m/z*) across different conditions to identify statistically significant difference [13]. Since it does not require metabolic labeling, dMS is a more versatile approach in biomarker discovery. dMS allows the comparison of multiple conditions simultaneously, such as longitudinal disease states, complex treatment conditions with different compounds, doses, time points, etc. Moreover, dMS takes into account biological variability since multiple replicates are used in each condition. The flexibility in comparing multiple conditions and the robustness in handling biological variability make label-free dMS a very powerful tool in protein biomarker discovery.

The label-free dMS approach has been used successfully to identify a number of biomarkers. For example, Paweletz and colleagues applied this platform to identify target engagement markers in cerebrospinal fluid (CSF) from rhesus monkeys treated with gamma secretase inhibitors (GSI) [14]. A total of 108 individual CSF samples from six monkeys, six time points and treatment with two potent GSI or vehicle in a 3-way crossover study, were analyzed by multifactorial dMS. By profiling of rhesus CSF samples, 39,567 features with distinct *m/z*, retention times, and intensity values, were extracted from the LC-MS data. Among them, 26 features were significantly altered in CSF after drug treatment. Remarkably, the relative abundance of these 26 features showed the same temporal profile as the ELISA measured levels of CSF Aβ 42 peptide, a known pharmacodynamic marker for γ-secretase inhibition. In addition, 20 features were identified by MS/MS as seven different peptides from CD99, a cell surface protein. These data demonstrate that dMS is a promising approach for the discovery, quantification, and identification of candidate biomarkers in CSF.

Contrasting the unbiased and open-ended MS-based proteomics profiling, antibody and protein microarrays are targeted approaches to identify protein biomarker candidates and to perform functional proteomic analyses. The antibody and protein microarray approach includes: (1) antibody (forward-phase) arrays with highly specific antibodies printed on chips or on uniquely identifiable beads, (2) purified peptides or proteins arrays for the detection of autoantibodies, (3) protein extracts arrays also used for the detection of autoantibodies [15]. One of the major advantages of targeted antibody/protein arrays is that they allow hypothesis-driven experimental designs in close-ended systems, such as the examination of key proteins in specific pathways as candidate biomarkers. Thus, the biological roles or functions of identified biomarkers are intimately associated with disease states or mechanism of action of drug treatment, leading to rational interpretation of identified biomarkers. For example, using reverse-phase protein microarray analysis of laser capture microdissected colorectal cancer tumor specimens, Pierobon and colleagues discovered that members of the EGFR and COX2 signaling pathways appear differentially activated in the primary tumors of patients with synchronous metastatic disease [16]. To characterize the differential expression of serum autoantibodies in Alzheimer's disease (AD) and non-demented control (NDC) groups, Nagele and colleagues utilized human protein microarrays, containing 9486 unique human protein antigens, and identified a set of 10 autoantibody biomarkers that can differentiate AD from NDC with 96.0 % sensitivity and 92.5 % specificity. Thus, serum autoantibodies can be used as highly specific and accurate biomarkers [17]. In between, SomaLogic (Boulder, CA) offers a mid-density targeted profiling approach for protein biomarker discovery, which could be considered to fall somewhere in between the MS-based profiling approach and the hypothesis-driven targeted antibody/protein array approach [18]. SOMAscan™ is an aptamer-based protein profiling platform which can measure 1129 proteins simultaneously from a wide variety of biological matrices. Ostroff and colleagues utilized SOMAscan™ to identify markers for early detection of lung cancer [19]. In a multi-center case-control study in archived serum samples from 1326 subjects with non-small cell lung cancer (NSCLC) in long-term tobacco-exposed populations, 44 candidate protein biomarkers were identified, and a 12-protein panel was developed that discriminates NSCLC from controls with 91 % sensitivity and 84 % specificity in cross-validated training and 89 % sensitivity and 83 % specificity in a separate verification test set, with similar performance for early and late stage NSCLC. Based on these findings, the investigators have initiated clinical validation studies of populations at risk for lung cancer.

After a protein biomarker is identified either from unbiased profiling or a targeted array platform, singleplex or multiplex targeted assays are developed for biomarker validation and clinical qualification studies. Ultimately, these targeted assays will be used for biomarker implementation in clinical development as well as for post-approval marketing and commercialization as companion diagnostics. These targeted assays include (1) immunochemistry-based assays such as ELISA, MSD, Luminex, Gyros, Singulex, etc., (2) MS-based assays such as selective ion monitoring (SRM) and multiplex reaction monitoring (MRM), and (3) aptamer-based protein quantification.

Metabolite Biomarker Discovery by Profiling

Downstream of genome, transcriptome, and proteome, metabolites are the real-time proximal reporters of alterations in the body in response to disease process. Metabolomics is the systematic and parallel analysis of many metabolites in a given tissue or biofluid. These metabolites are low molecular weight compounds including sugars, amino acids, organic acids, nucleotides and lipids. Like the approaches in proteomics, metabolite markers can be identified either by open-ended profiling to measure as many metabolites as possible or by targeted methods to focus on a particular class of compounds, typically several dozen to hundreds of metabolites. Both nuclear magnetic resonance spectroscopy (NMR) and MS can be used to analyze metabolites in a non-targeted, pattern recognition manner or a targeted manner. These approaches have identified metabolite markers for early diagnosis of acute kidney injury, polycystic kidney disease, and kidney cancer, and metabolite markers associated with cardiovascular disease and heart disease [20, 21].

Other technologies that may be used in biomarker discovery include flow cytometry and immunohistochemistry. Flow cytometry has the ability to sort cells from complex biological matrices and to measure analytes on the surface of cells or inside them. Novel technologies like mass cytometry (e.g., CyTOF®), combine the single-cell selectivity of flow cytometry with the exquisite selectivity of mass spectrometry to enable simultaneous measurements of dozens of markers of cell types without interference from spectral overlap or autofluorescence. This technology may have a major impact on the identification of biomarkers, especially in oncology and inflammatory diseases [22, 23].

From Discovery to the Clinic

The naturally desired progression for candidate biomarkers is the transition from bench to clinic. Before that takes place, biomarkers need to be subjected to analytical validation and clinical qualification. Analytical validation includes the assessment of the effect of pre-analytical variability, sensitivity, intra- and inter-assay precision, accuracy, specificity, matrix effects, biological variability, analyte stability, and identification of possible interfering substances. The analytical validation could take a few weeks to complete, whereas biomarker qualification may take many years to be accomplished. Biomarker qualification refers to the clinical validation of a biomarker—that is, the evidentiary process of linking a biomarker with biological/ pathological processes and clinical endpoints [24]. Biomarker qualification is also a fit-for-purpose in the sense that the level of qualification is dictated by the intended use of the biomarker. Wagner [24] proposed a graded classification for the biomarker qualification process. This extends from exploration, demonstration and characterization biomarkers to surrogacy, in which the biomarker can substitute for a clinical endpoint. This proposed graded process captures the utility of the biomarkers not only for decision-making but also for regulatory purposes. In summary, the process from biomarker discovery to their wide implementation into the clinic could be long and resource-intensive. In order to minimize time and cost, biomarker researches ought to always keep the fit-or-purpose concept in mind, not only in the analytical aspects but also in their development and implementation in the clinic.

References

1. Biomarkers and surrogate endpoints (2001) preferred definitions and conceptual framework. Clin Pharmacol Ther 69(3):89–95
2. Laterza OF, Hendrickson RC, Wagner JA (2007) Molecular biomarkers. Drug Inf J 41 (5):573–585
3. van't Veer LJ, Dai H, van de Vijver MJ et al (2002) Gene expression profiling predicts clinical outcome of breast cancer. Nature 415(6871):530–536
4. van de Vijver MJ, He YD, van't Veer LJ et al (2002) A gene-expression signature as a predictor of survival in breast cancer. N Engl J Med 347(25):1999–2009
5. Hetherington S, Hughes AR, Mosteller M et al (2002) Genetic variations in HLA-B region and hypersensitivity reactions to abacavir. Lancet 359(9312):1121–1122
6. Narod SA, Foulkes WD (2004) BRCA1 and BRCA2: 1994 and beyond. Nat Rev Cancer 4 (9):665–676

7. Laterza OF, Lim L, Garrett-Engele PW et al (2009) Plasma MicroRNAs as sensitive and specific biomarkers of tissue injury. Clin Chem 55(11):1977–1983
8. Laterza OF, Scott MG, Garrett-Engele PW, Korenblat KM, Lockwood CM (2013) Circulating miR-122 as a potential biomarker of liver disease. Biomark Med 7(2):205–210
9. Tsongalis GJ, Lockwood CM (2013) Clinical utility of miRNAs in diagnosis and prognosis. Clin Biochem 46(10–11):839
10. Laterza OF, Modur VR, Crimmins DL et al (2006) Identification of novel brain biomarkers. Clin Chem 52(9):1713–1721
11. Cox J, Mann M (2011) Quantitative, high-resolution proteomics for data-driven systems biology. Annu Rev Biochem 80:273–299
12. Gygi SP, Rist B, Gerber SA, Turecek F, Gelb MH, Aebersold R (1999) Quantitative analysis of complex protein mixtures using isotope-coded affinity tags. Nat Biotechnol 17(10):994–999
13. Wiener MC, Sachs JR, Deyanova EG, Yates NA (2004) Differential mass spectrometry: a label-free LC-MS method for finding significant differences in complex peptide and protein mixtures. Anal Chem 76(20):6085–6096
14. Paweletz CP, Wiener MC, Bondarenko AY et al (2010) Application of an end-to-end biomarker discovery platform to identify target engagement markers in cerebrospinal fluid by high resolution differential mass spectrometry. J Proteome Res 9(3):1392–1401
15. Sanchez-Carbayo M (2011) Antibody microarrays as tools for biomarker discovery. Methods Mol Biol 785:159–182
16. Pierobon M, Calvert V, Belluco C et al (2009) Multiplexed cell signaling analysis of metastatic and nonmetastatic colorectal cancer reveals COX2-EGFR signaling activation as a potential prognostic pathway biomarker. Clin Colorectal Cancer 8(2):110–117
17. Nagele E, Han M, DeMarshall C, Belinka B, Nagele R (2011) Diagnosis of Alzheimer's disease based on disease-specific autoantibody profiles in human sera. PLoS One 6(8):e23112
18. Gold L, Ayers D, Bertino J et al (2010) Aptamer-based multiplexed proteomic technology for biomarker discovery. PLoS One 5(12):e15004
19. Ostroff RM, Bigbee WL, Franklin W et al (2010) Unlocking biomarker discovery: large scale application of aptamer proteomic technology for early detection of lung cancer. PLoS One 5 (12):e15003
20. Rhee EP, Gerszten RE (2012) Metabolomics and cardiovascular biomarker discovery. Clin Chem 58(1):139–147
21. Weiss RH, Kim K (2012) Metabolomics in the study of kidney diseases. Nat Rev Nephrol 8 (1):22–33
22. Newell EW, Sigal N, Bendall SC, Nolan GP, Davis MM (2012) Cytometry by time-of-flight shows combinatorial cytokine expression and virus-specific cell niches within a continuum of CD8+T cell phenotypes. Immunity 36(1):142–152
23. Bendall SC, Nolan GP (2012) From single cells to deep phenotypes in cancer. Nat Biotechnol 30(7):639–647
24. Wagner JA (2008) Strategic approach to fit-for-purpose biomarkers in drug development. Annu Rev Pharmacol Toxicol 48:631–651

Application of Quantitative Biomeasures in Early Drug Discovery

Scott T. Fountain and Paolo Vicini

Abstract Mathematical models of drug action are essential in contemporary drug discovery and development. Applications include exposure-response modeling (pharmacokinetics-pharmacodynamics or PK-PD); quantitative understanding of biological target and pathway; and systems approaches that integrate characteristics of the biological system with associated drug exposure. Encompassing empirical, mechanistic, or semi-mechanistic approaches, these mathematical models are informed by experimental data quantifying not only drug exposure (pharmacokinetics) and associated biological response (biomarkers), but also system-specific parameters intermediate between drug exposure and response. These system-specific endpoints, or biomeasures, include target-specific measurements such as density, turnover, shedding, and internalization rate. Quantifying these pharmacokinetic and pharmacodynamic endpoints—which include small molecule, biological, and cellular measures—requires a diverse repertoire of analytical instrumentation and approaches. The discipline partnership between quantitative bioanalytics and systems modeling provides an invaluable tool to improve the success of pharmaceutical research and development. The authors will provide a perspective on the interface between laboratory science and mathematical modeling to improve assessment of exposure-response relationships, and ultimately successful drug development.

Keywords Biomeasure · Biomarker · Target · Exposure-Response · PK-PD · Systems pharmacology · Systems models · Quantitative

---

S.T. Fountain (✉) · P. Vicini
Pharmacokinetics, Dynamics and Metabolism—New Biological Entities,
Pfizer Worldwide Research and Development, 10777 Science Center Drive,
San Diego, CA 92121, USA
e-mail: scott@stfountain.com

S.T. Fountain
Present Address: Encinitas, CA, USA

P. Vicini
Present Address: MedImmune, Granta Park, Cambridge, UK

R. Weiner and M. Kelley (eds.), *Translating Molecular Biomarkers into Clinical Assays*, AAPS Advances in the Pharmaceutical Sciences Series 21,
DOI 10.1007/978-3-319-40793-7_4

Introduction: Quantitative Methods in Translational Research

Quantitative understanding of biological targets and pathways is ever more important in modern drug discovery and development, where the tolerance for risk is low and R&D productivity unsustainable [1]. Recently, it has been demonstrated how understanding drug exposure at the desired site of action, associated target engagement, and subsequent disease modulation (the *Three Pillars of Survival*) greatly contributes to de-risking drug discovery and development programs, particularly where direct measurement and quantification of exposure-response endpoints is obtained [2]. Program success appears related to the rigorous assessment of well-defined endpoints resulting from specific biological queries. These seemingly simple queries include: Will the drug get to where it is supposed to go (*is there exposure at the biological target*)? Does anything happen to the intended target when the drug is present (*is there target binding*)? Once that happens, is the target modulated as would be expected (*is there pharmacological activity*)? Though it is recognized that direct measurement of exposure, binding, and pharmacology are not sufficient to guarantee robust inference of drug action per se, application of quantitative bioanalysis of well-defined endpoints in support of this goal is necessary.

The drug discovery and development paradigm has made great strides, evolving from a largely empirical discipline to one increasingly driven by predictive approaches and science. Given the complexity of modern drug programs, and especially the pathways they are intended to modulate, use of mathematical models to interpret data, predict outcomes, and design experiments is becoming paramount. Pharmacokinetics-pharmacodynamics (PK-PD), or exposure-response modeling, both in the translational [3] and the clinical setting [4], are exemplary tools that have reshaped drug discovery into a predictive science. Mathematical PK-PD models have been extensively used in drug development to interpret available data, test mechanistic hypotheses or, at best, design experiments prospectively [5, 6]; a natural evolution of PK-PD is the emerging science of mechanistic, or systems, modeling [7].

Systems approaches, as applied in drug discovery and development, integrate aspects of a biological system with understanding of the drug's exposure (i.e., concentration), the ultimate goal being to link preclinical and clinical environments. When this is successful, compartmental models of the drug concentration profile associated with a certain dosing scheme can be integrated with subsequent target modulation or changes in relevant biomarkers, in a causal cascade that allows the researcher or clinician to propose a mechanism of action [8]. Requiring postulation of a causal relationship, models can be crafted and tested against existing data sets or, once validated, used prospectively, such as when data sets are limited or sparse [9]. The detail with which such causal relationships can be represented, or "modeled," is a function of the data informing the models, and will therefore dictate the required characteristics of biomeasure and biomarker data (e.g., assessments of

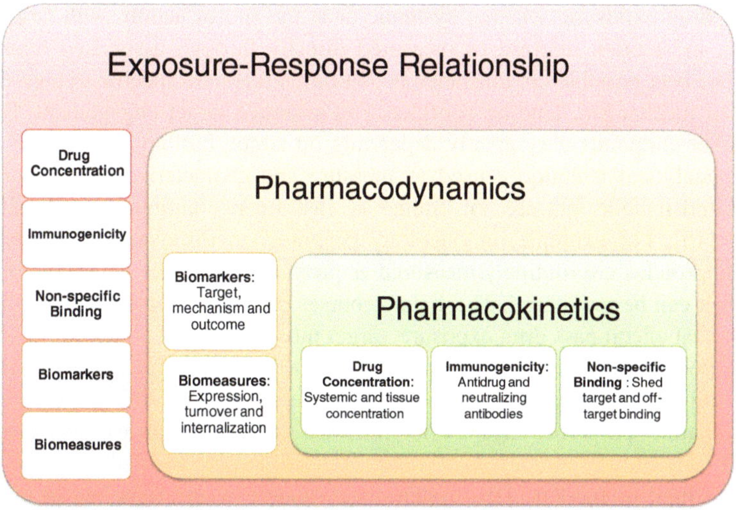

Fig. 1 Accumulated knowledge of in vivo and in vitro exposure-response relationships for investigative compounds is the result of nested interactions between understanding of the agent's pharmacokinetics (exposure) and pharmacodynamics (response). While this applies to small and large molecules and in general holds across modalities, some features of exposure-response (e.g., immunogenicity) are specific to biotherapeutics

selectivity, sensitivity, number of variables, etc.) for the model to produce valuable predictions. Clearly, with increasing detail in the system's mathematical representation, equally detailed experimental information is necessary to drive model building and incorporate the appropriate level of mechanistic detail. As such, models can be empirical [10], mechanistic or semi-mechanistic [11], or include richly detailed system-level factors [12]. We will describe how laboratory science and mathematical modeling interface, and how one influences the other to address the challenges pharmaceutical research and development is facing [1], specifically the granular understanding of the in vitro and in vivo drivers of exposure-response (Fig. 1).

We begin with a brief review of how PK-PD and system-level models have been used in drug discovery and development, after which we will move to a description of how these concepts can be integrated with quantitative bioanalysis to improve our understanding of how key targets and pathways can be modulated through pharmacological intervention.

Value and Application of PK-PD and Systems Models in Drug Discovery

As mentioned, mathematical models of drug action have a distinguished history and continue to evolve [13]. Most often these models address PK-PD relationships, linking drug exposure, whether systemic or at the site of action, with target modulation, most often inferred or measured directly through associated biomarkers [14]. The best possible circumstance is that biomarkers are specific to a given drug target or mechanism, lending confidence in selective target engagement. Because biomarker endpoints vary greatly depending on target, biology, and site of action, the bioanalytical techniques used to measure and characterize them are equally varied and include, but are not limited to, instrument-, antibody-, and cell-based assays [15]. For example, in oncology programs to discover kinase inhibitors, phosphoproteins are routinely measured as proximal substrates where inhibition of the target can be monitored. These time courses can then be integrated into PK-PD models that encompass drug exposure, target modulation and antitumor effect [16]. These models are increasingly popular and can be used to characterize maximal effect and half-maximal effect exposure, quantitatively describing a drug's pharmacodynamic properties against its intended target. Again we recognize that PK-PD models, despite all their value, are still best classified as empirical or semi-mechanistic models. True system-level models, where signaling or metabolic pathways are explicitly represented in their constituent parts, remain comparatively rare [8]. Their emergence is arguably the next frontier for applied modeling and simulation in pharmaceutical research. The emerging discipline of systems pharmacology offers promise for drug discovery especially when it is symbiotically linked to available laboratory methods that provide unprecedented quantification of in vitro and in vivo biology [17]. Systems pharmacology is a unique evolution of systems biology in the following ways [12]:

1. Systems pharmacology recognizes the importance of time-dependent data and time series, thus using differential as opposed to algebraic or steady-state equations;
2. Systems pharmacology uses drugs and pharmacologic agents to probe the system and investigate how it responds, thus focusing on dynamic changes, as opposed to homeostatic behavior;
3. Systems pharmacology uses tools defining pathway analysis, i.e., PK-PD and systems biology, to ultimately predict the in vivo behavior of intact systems and their response to a variety of perturbations.

Because it is well differentiated from other modeling approaches not only in terms of complexity, but also in terms of intended impact and use, systems pharmacology has generated increasing interest from both academia and industry [18]. This requires parallel advancements in quantitative bioanalytical platforms and methodologies, necessary to inform these more complex mathematical models. Convergence of these independent disciplines, quantitative bioanalytics and systems models, has the potential to create a differentiated toolkit for pharmaceutical research—a truly quantitative approach to the understanding of biological targets and their pathways—building on the inherent reciprocity of these sciences. Specifically, this can inform a *virtuous circle*, i.e., how establishment of systems pharmacology models can motivate bioanalysis, and vice versa.

Data Requirements to Inform Quantitative and Systems Pharmacology Models

In principle, a well-characterized mathematical model or prediction should describe and reconcile appropriately quantified biological variables. Data requirements vary depending on the model's intended purpose, for example, whether a mathematical model is used to make pharmacokinetic predictions, which can be validated against readily obtainable data [19], or pharmacodynamic scaling [20], where efficacy predictions can be made using nonclinical models, then compared to clinical observations. In the clinic, given patient availability and logistical constraints, measurements are often limited to drug concentrations. There is, however, the potential for quantitative bioanalysis to provide essential mechanistic parameters for systems pharmacology models, whose predictions can then be iteratively tested and updated when new experimental evidence comes to light. The concept of "*biomeasure*" is an example of successful synergy between bioanalysis and modeling and simulation. Biomeasures can be defined as drug-independent characteristics of the biological system, such as receptor density or target turnover, necessary to successfully implement mechanistic, predictive models [21].

What Are Biomeasures? Their Context and Application

A Brief Introduction to Biomeasures

Biomeasures are a relatively new concept finding favor in applied drug discovery [13]. Biomeasures are system-specific parameters that are intermediate between drug exposure (measured by concentrations) and its response (quantified through biomarkers). Examples of biomeasures in a drug discovery program may include target density, target turnover, target shedding, and rate of internalization. Biomeasures are necessary to fully characterize how drug molecules trigger the mechanistic cascade that ultimately leads to effect. Depending on how much detail is needed to make informed predictions on the system, the number of biomeasures required to inform a model can be large or small, and the choice of what biomeasures to monitor depends on the priorities of a particular program. It is important to note this is not a one-size-fits-all approach. Differences in target properties, e.g., turnover, can motivate different approaches in how quantitative bioanalysis is performed and prioritized, especially in large molecule development [22].

Key Biomeasures of Interest in Drug Development

Depending on how the drug interacts with the target and how complex the pathway of interest is, a varying amount of detail is necessary to elucidate the therapy's mechanism of action [23]. This is the role played by emerging tools such as proteomics [24] and techniques to measure the immune response [25] that may, for example, be required to obtain a mechanistic understanding of immunogenicity [26].

PK-PD methods have a long and illustrious history in facilitating the understanding of reversible inhibitors' mechanisms of action [27]. An example of where target properties become crucial even for small molecule drugs is in irreversible inhibitors, valuable in treating various diseases [28]. The development of irreversible inhibitors is relatively recent [29] and in this therapeutic class target turnover plays a major role in drug target engagement and pharmacology, as shown in [30], thus indicating turnover as a key biomeasure to assess in a comprehensive model.

A Review of Established and Emerging Tools Informing PK-PD and System Models

The predictive quality of exposure-response models relies heavily on the quality of data used to inform the model. Data quality itself—accuracy, precision, and specificity, for instance—is a function of analytical methodology or platform, access to reliable reference standards and controls, and understanding of the biological system itself, among other variables. A modeler can obtain data from a number of sources. Often data on target expression and internalization may be published in the literature, and text mining [31] may provide biomeasure data sufficiently robust to initially inform a model in early development or for biomeasure endpoints that are particularly difficult or impractical to obtain. Target expression and related biomeasure data may also be available from biological studies that characterize protein expressing using Western blot or immunohistochemistry (IHC). However, in order to provide reliable data that comparably informs both "halves" of the PK-PD continuum, so to speak, bioanalytical approaches that independently quantify exposure, response, and target assessment, applying more rigorous analytical tools, may be favored or required. This is particularly relevant in systems where the predictive strength of the model necessitates higher precision, accuracy, or specificity. "Stress points" in the model, obtained via one of the many flavors of sensitivity analysis [32], can highlight the need to quantify a specific flux or control parameter most accurately, or, conversely, can suggest that accurate quantification is not crucial for model predictions. Such approaches can provide a quantitative basis to set laboratory objectives for a new discovery program (potency, target engagement mechanism, extent of target modulation desired, etc.).

Assessing Drug Exposure—PK

In order to make quality drug exposure measurements, well-characterized standards and biological matrix free of endogenous analyte is critical. Established bioanalytical platforms such as mass spectrometry and antibody-based approaches such as enzyme-linked immunosorbent assays (ELISA), arguably the workhorse methodologies of the contemporary bioanalytical pharmacokinetic laboratory, are well suited for these bioanalytical applications, and acceptance criteria for assay performance have been established and applied for pharmacokinetic, bioavailability, and bioequivalence assessment for a number of years [33]. Additional analytical capabilities inherent in mass spectrometry and antibody-based analytics include multiplexing [34, 35], automation, including data interchange [36], and common platform expertise allowing assay transfer between laboratories. These characteristics of established PK bioanalysis are worth noting, as they have arguably become the benchmark used to assess the quality of non-PK bioanalytical methods.

Assessing Drug Effect and Target Engagement—PD, Biomarkers and Biomeasures

In comparison to therapeutic drug assessment, biomarkers and biomeasures are, by definition, endogenous endpoints and are often not fully characterized, particularly since they are often macromolecules that exist in multiple isoforms or chemically modified states. This adds complexity and ambiguity to the analysis. To address the inherent ambiguity in biomarker data obtained by various assays, Lee et al., have recommended a system to categorize biomarker assay data based on the type of assay employed [37]. This nomenclature defines biomarker assays as (a) definitive quantitative, (b) relative quantitative, (c) quasi-quantitative, and (d) qualitative, reflecting variability in access to, or purity of, definitive reference standards and specifics of experimental design. This scaled approach characterizing the quantitative rigor of endogenous analytes provides a convenient framework to recognize the quantitative limits of the analytical methodology, mitigating the risk of over-interpretation of model projections or over-interpretation of model estimates. This is of equal relevance when investigating a molecule's safety and efficacy [38].

Emerging and Innovative Tools

Innovative bioanalytical tools are evolving with unique capabilities in addressing biomeasure endpoints, including target expression, turnover, and internalization. Two of these approaches, imaging flow cytometry and mass cytometry, build on the flow cytometry platform, which, like mass spectrometry and antibody-based methods, is a cornerstone of the contemporary pharmaceutical laboratory, as shown by published applications addressing mechanism-based assessment of target engagement and safety [39]. Imaging flow cytometry builds on the capabilities of flow cytometry, adding spatially separated imaging and digital microscopy [40] that provides unique capabilities in assessing biomarkers and biomeasures [41]. Mass cytometry couples fluorescent-based flow cytometry with inductively coupled mass spectrometry to quantify epitope-specific antibodies custom labeled with rare earth isotopes, providing unparalleled multiplexing capabilities in assessing surface antigen expression [42, 43]. These and other tools promise to provide differentiated improvements, provided interpretative models continue their symbiotic relationship with laboratory sciences, and vice versa.

Acknowledgments The Authors are grateful to Dr. Bill Smith for his comments on an earlier version of this chapter.

References

1. Scannell JW, Blanckley A, Boldon H, Warrington B (2012) Diagnosing the decline in pharmaceutical R&D efficiency. Nat Rev Drug Discov 11:191–200
2. Morgan P, Van Der Graaf PH, Arrowsmith J, Feltner DE, Drummond KS, Wegner CD, Street SDA (2012) Can the flow of medicines be improved? Fundamental pharmacokinetic and pharmacological principles toward improving Phase II survival. Drug Discov Today 17:419–424
3. Van Der Graaf PH, Gabrielsson J (2009) Pharmacokinetic-pharmacodynamic reasoning in drug discovery and early development. Future Med Chem 1:1371–1374
4. Peck CC (2010) Quantitative clinical pharmacology is transforming drug regulation. J Pharmacokinet Pharmacodyn 37:617–628
5. Lalonde RL, Kowalski KG, Hutmacher MM, Ewy W, Nichols DJ, Milligan PA, Corrigan BW, Lockwood PA, Marshall SA, Benincosa LJ, Tensfeldt TG, Parivar K, Amantea M, Glue P, Koide H, Miller R (2007) Model-based drug development. Clin Pharmacol Ther 82:21–32
6. Milligan PA, Brown MJ, Marchant B, Martin SW, van der Graaf PH, Benson N, Nucci G, Nichols DJ, Boyd RA, Mandema JW, Krishnaswami S, Zwillich S, Gruben D, Anziano RJ, Stock TC, Lalonde RL (2013) Model-based drug development: a rational approach to efficiently accelerate drug development. Clin Pharmacol Ther 93:502–514
7. Danhof M, de Jongh J, De Lange ECM, Della Pasqua O, Ploeger BA, Voskuyl RA (2007) Mechanism-based pharmacokinetic-pharmacodynamic modeling: biophase distribution, receptor theory, and dynamical systems analysis. Annu Rev Pharmacol Toxicol 47:357–400
8. Kirouac DC, Onsum MD (2013) Using network biology to bridge pharmacokinetics and pharmacodynamics in oncology, CPT. Pharmacomet Syst Pharmacol 2:e71
9. Tibbitts J, Cavagnaro JA, Haller CA, Marafino B, Andrews PA, Sullivan JT (2010) Practical approaches to dose selection for first-in-human clinical trials with novel biopharmaceuticals. Regul Toxicol Pharmacol 58:243–251
10. Gabrielsson J, Fjellstrom O, Ulander J, Rowley M, Van Der Graaf P (2011) Pharmacodynamic-pharmacokinetic integration as a guide to medicinal chemistry. Curr Top Med Chem 11:404–418
11. Mager DE (2006) Target-mediated drug disposition and dynamics. Biochem Pharmacol 72:1–10

12. Iyengar R, Zhao S, Chung S-W, Mager DE, Gallo JM (2012) Merging systems biology with pharmacodynamics. Sci Transl Med 4:126ps127
13. Agoram BM, van der Graaf PH (2012) Biomarkers and biomeasures: key enablers for pharmacokinetic–pharmacodynamic modeling in drug discovery and development. Bioanalysis 4:1143–1145
14. Atkinson AJ, Colburn WA, DeGruttola VG, DeMets DL, Downing GJ, Hoth DF, Oates JA, Peck CC, Schooley RT, Spilker BA, Woodcock J, Zeger SL (2001) Biomarkers and surrogate endpoints: preferred definitions and conceptual framework. Clin Pharmacol Ther 69:89–95
15. Lee J, Devanarayan V, Barrett Y, Weiner R, Allinson J, Fountain S, Keller S, Weinryb I, Green M, Duan L, Rogers J, Millham R, O'Brien P, Sailstad J, Khan M, Ray C, Wagner J (2006) Fit-for-Purpose method development and validation for successful biomarker measurement. Pharm Res 23:312–328
16. Wang S, Guo P, Wang X, Zhou Q, Gallo JM (2008) Preclinical pharmacokinetic/pharmacodynamic models of gefitinib and the design of equivalent dosing regimens in EGFR wild-type and mutant tumor models. Mol Cancer Ther 7:407–417
17. Van der Graaf PH (2012) CPT: pharmacometrics and systems pharmacology. Pharmacomet Syst Pharmacol 1:e8
18. Sorger PK, Allerheiligen SRB, Abernethy DR, Altman RB, Brouwer KLR, Califano A, D'Argenio DZ, Iyengar R, Jusko WJ, Lalonde R, Lauffenburger DA, Shoichet B, Stevens JL, Subramaniam S, Van der Graaf PH, Vicini P (2011) Quantitative and systems pharmacology in the post-genomic era: new approaches to discovering drugs and understanding therapeutic mechanisms, An NIH White Paper by the QSP Workshop Group, p 48
19. Luu KT, Bergqvist S, Chen E, Hu-Lowe D, Kraynov E (2012) A model-based approach to predicting the human pharmacokinetics of a monoclonal antibody exhibiting target-mediated drug disposition. J Pharmacol Exp Ther 341:702–708
20. Yamazaki S, Vicini P, Shen Z, Zou HY, Lee J, Li Q, Christensen JG, Smith BJ, Shetty B (2012) Pharmacokinetic/pharmacodynamic modeling of crizotinib for anaplastic lymphoma kinase inhibition and antitumor efficacy in human tumor xenograft mouse models. J Pharmacol Exp Ther 340:549–557
21. van der Graaf P, Benson N (2011) Systems pharmacology: bridging systems biology and pharmacokinetics-pharmacodynamics (PKPD) in drug discovery and development. Pharm Res 28:1460–1464
22. Baumann A (2008) Preclinical development of therapeutic biologics. Expert Opin Drug Discov 3:289–297
23. Xu K, Liu L, Saad OM, Baudys J, Williams L, Leipold D, Shen B, Raab H, Junutula JR, Kim A, Kaur S (2011) Characterization of intact antibody–drug conjugates from plasma/serum in vivo by affinity capture capillary liquid chromatography–mass spectrometry. Anal Biochem 412:56–66
24. Krüger M, Moser M, Ussar S, Thievessen I, Luber CA, Forner F, Schmidt S, Zanivan S, Fässler R, Mann M (2008) SILAC mouse for quantitative proteomics uncovers kindlin-3 as an essential factor for red blood cell function. Cell 134:353–364
25. Bendall SC, Simonds EF, Qiu P, Amir E-AD, Krutzik PO, Finck R, Bruggner RV, Melamed R, Trejo A, Ornatsky OI, Balderas RS, Plevritis SK, Sachs K, Pe'er D, Tanner SD, Nolan GP (2011) Single-cell mass cytometry of differential immune and drug responses across a human hematopoietic continuum. Science 332:687–696
26. Chen X, Hickling T, Kraynov E, Kuang B, Parng C, Vicini P (2013) A mathematical model of the effect of immunogenicity on therapeutic protein pharmacokinetics. AAPS J 15:1141–1154
27. Mager DE, Wyska E, Jusko WJ (2003) Diversity of mechanism-based pharmacodynamic models. Drug Metab Dispos 31:510–518
28. Wang Z, Watt W, Brooks NA, Harris MS, Urban J, Boatman D, McMillan M, Kahn M, Heinrikson RL, Finzel BC, Wittwer AJ, Blinn J, Kamtekar S, Tomasselli AG (2010) Kinetic and structural characterization of caspase-3 and caspase-8 inhibition by a novel class of irreversible inhibitors. Biochimica Biophysica Acta (BBA) Proteins Proteomics 1804:1817–1831

29. Kim K-H, Maderna A, Schnute ME, Hegen M, Mohan S, Miyashiro J, Lin L, Li E, Keegan S, Lussier J, Wrocklage C, Nickerson-Nutter CL, Wittwer AJ, Soutter H, Caspers N, Han S, Kurumbail R, Dunussi-Joannopoulos K, Douhan Iii J, Wissner A (2011) Imidazo[1,5-a] quinoxalines as irreversible BTK inhibitors for the treatment of rheumatoid arthritis. Bioorg Med Chem Lett 21:6258–6263
30. Barf T, Kaptein A (2012) Irreversible protein kinase inhibitors: balancing the benefits and risks. J Med Chem 55:6243–6262
31. Korhonen A, Séaghdha DÓ, Silins I, Sun L, Högberg J, Stenius U (2012) Text mining for literature review and knowledge discovery in cancer risk assessment and research. PLoS ONE 7:e33427
32. Shoemaker JE, Doyle FJ III (2008) Identifying fragilities in biochemical networks: robust performance analysis of fas signaling-induced apoptosis. Biophys J 95:2610–2623
33. Smith G (2010) Bioanalytical method validation: notable points in the 2009 draft EMA Guideline and differences with the 2001 FDA Guidance. Bioanalysis 2:929–935
34. Ellington AA, Kullo IJ, Bailey KR, Klee GG (2010) Antibody-based protein multiplex platforms: technical and operational challenges. Clin Chem 56:186–193
35. Hale JE (2013) Advantageous uses of mass spectrometry for the quantification of proteins. Int J Proteomics 2013:6
36. Leung S, Usansky J, Lynde R, Thway T, Hendricks R, Rusnak D (2012) Ligand-binding assays in the 21st century laboratory: recommendations for an automated data interchange process. AAPS J 14:105–112
37. Lee JW, Smith WC, Nordblom GD, Bowsher RR (2003) Biomarkers in clinical drug development. In: Bloom C, Dean RA (eds) Marcel Dekker, New York
38. Hansel TT, Kropshofer H, Singer T, Mitchell JA, George AJT (2010) The safety and side effects of monoclonal antibodies. Nat Rev Drug Discov 9:325–338
39. Henriques CM, Rino J, Nibbs RJ, Graham GJ, Barata JT (2010) IL-7 induces rapid clathrin-mediated internalization and JAK3-dependent degradation of IL-7Rα in T cells. Blood 115:3269–3277
40. Basiji DA, Ortyn WE, Liang L, Venkatachalam V, Morrissey P (2007) Cellular image analysis and imaging by flow cytometry. Clin Lab Med 27:653–670
41. Zhu M, Pleasic-Williams S, Lin T, Wunderlich D, Cheng J, Masferrer J (2013) pSTAT3: a target biomarker to study the pharmacology of the anti-IL-21R antibody ATR-107 in human whole blood. J Transl Med 11:65
42. Janes MR, Rommel C (2011) Next-generation flow cytometry. Nat Biotech 29:602–604
43. Bodenmiller B, Zunder ER, Finck R, Chen TJ, Savig ES, Bruggner RV, Simonds EF, Bendall SC, Sachs K, Krutzik PO, Nolan GP (2012) Multiplexed mass cytometry profiling of cellular states perturbed by small-molecule regulators. Nat Biotech 30:858–867

Biomarkers in Discovery and Preclinical Phase During Drug Development

Jean W. Lee

Abstract Knowledge of complex diseases and biological pathways of disease progression contributes to the identification of drug targets and mechanistic drug development of therapeutics that act on specific targets. Preclinical data on toxicokinetics, pharmacokinetics, and pharmacodynamics from appropriate animal species are necessary for drug candidate selection and model building to determine the first time in human starting dose and clinical study design. This chapter discusses the selection of animal models with biological and toxicological relevancy to humans, using pathway biomarkers to inform target hit, target engagement, and off-target effects. The preclinical studies allow determinations of biomarkers from specimens of readily accessible matrices and internal organs/tissues to confirm the applicability of efficacious and safety biomarkers.

Keywords Preclinical PK/PD · Pathway biomarkers · Drug candidate selection · Target engagement

---

Financial and competing interests disclosure

The authors have no relevant affiliations or financial involvement with any organization or entity with a financial interest in or financial conflict with the subject matter or materials discussed in the manuscript. This includes employment, consultancies, honoraria, stock ownership or options, expert testimony, grants or patents received or pending, or royalties.

No writing assistance was utilized in the production of this manuscript.

J.W. Lee (✉)
BioQualQuan, 370 Spring Park Road, Camarillo, CA 93012, USA
e-mail: ljean095@gmail.com

R. Weiner and M. Kelley (eds.), *Translating Molecular Biomarkers into Clinical Assays*, AAPS Advances in the Pharmaceutical Sciences Series 21,
DOI 10.1007/978-3-319-40793-7_5

Overview: The Necessary Application of Biomarkers in Animal Models During Drug Development

Genomics and proteomics information have increased our understanding on the biology of disease initiation and progression. Knowledge of complex diseases and the biological pathways of disease progression contribute to systems biology. New drug targets have emerged through systems biology, resulting in specific, rational drug development of therapeutics. During the discovery phase a potential target is identified from genomics and proteomics information from in vitro cell biology and epidemiology. Impactful agents of small molecule (SM) and protein are created to test the hypothesis of the targeted effect on the disease. Multitudes of candidates are produced from combinatorial chemistry for SM and from recombinant technology for large molecule (LM). The target molecules can be used for the initial in vitro screening of binding affinity and cell based activity using human cell lines. Animal models are necessary to bring about the understanding of target hit, target engagement and the establishment of dose exposure and effect relationship of toxicokinetics (TK), pharmacokinetics (PK) and pharmacodynamics (PD). The interface of systems biology and TK/PK/PD is the basis of systems pharmacology to effectively guide drug development [1, 2].

Clinical trials are increasingly more extensive and complex, with probable failures due to efficacy, safety, or commercial concerns. To decrease the risk of high failure rates, data from animal studies have been used in TK/PK/PD modeling to predict efficacy and safety. These data draw from reliable determinations of the target biomarker and downstream effect biomarkers of positive (efficacy) and negative (safety) effects.

At least one animal species that bears resemblance in the diseases biology to that of humans should be used. Because of the varieties of biomarkers (target, efficacious PD, and safety), the TK/PK/PD data interpretation and modeling can be challenging and multiple animal models may be required. Several appropriate animal species can be selected based upon the similar aspects of the target biological pathways and toxicity responses.

The biomarkers should be present in an accessible matrix that represents the disease target organ or tissue in order to be measured. A biomarker assay strategy for the animal species and selected biomarkers should be developed with the intended purposes of using the preclinical TK/PK/PD data in model building to aid the FIH starting dose decision and clinical study design. In this chapter, the biological relevancy of animal models is discussed, followed by the determinations of efficacy and safety biomarkers in animal species to assess benefit/risk to aid go/no-go decision.

Biological Relevancy of Animal Models—Opportunities and Limitations of Animal Models

Traditionally, basic PK metabolism of SMs is characterized in rodents and nonhuman primates. Target-specific drug development has added sophistication to conventional preclinical studies: The primary consideration is the comparison of the disease pathway in the animal species to that of humans. Both similarities and differences must be carefully studied to establish the relevancy of the animal models. Biomarkers reflecting various biological pathways and drug effects are often used for comparison. Pathological phenomenon similar to that in human can be induced in natural species of rodents and nonhuman primates to build animal models. However, toxic responses may differ between human and animal species as in the case of the TGN1412 anti-CD28 molecule where cytokine responses in cynomolgous monkeys failed to predict the exaggerated immune stimulation (cytokine storm) in a phase I clinical trial [3, 4].

Xenografts, transgenics and knockouts which include some aspects of the human pathological environment have been produced as animal models to investigate specific cause-and-effect relationships of different disease impacting factors. Xenografts have been widely used to study cancer pathogenesis, progression and drug intervention. Human tumor cells are transplanted into immune-compromised mice, such as athymic nude mice and severely compromised immunodeficient (SCID) mice which do not reject the transplant. Transgenic and knockout mouse models use genetically engineered mice (GEM) with putative oncogenes that are mutated, over-expressed, or deleted. To mimic the human tissue microenvironment, stroma from the human tumor can be included in the xenograft transplant, and humanized mice can be used for GEMs or xenografts [5]. A variety of transgenic and knockout mouse models can be used to reflect the diversity in diseases with complex pathways and interwoven biological effects as in the example of investigations on various disease mechanisms of different human diabetic patients [6, 7].

Efficacy Biomarkers

Target Hit and Target Engagement

The target molecule for disease modulation is used for in vitro screening and then for in vivo tests of target hit and target engagement in animal models. Target hit of the drug is indicated by the binding of the drug candidate to the specific biochemical. Target engagement includes the events of target hit and modulations of downstream biological reactions and subsequent physiological responses. Useful biomarkers are selected based on the changes of the biochemical or physiological responses that can be objectively measured and bear a dose-dependent relationship to the drug exposure.

For confirmation of the in vitro screening results on target hit, modulation on the free target concentrations should be shown to be related to drug exposure in the animal model over a time course. The assessment of target hit depends on the target location:

(A) In blood circulation—Changes of a soluble target biomarker after dosing can be directly measured in the blood. For drugs with antagonistic action, suppression of the unbound target can be shown. For example, serum concentrations of free tumor necrosis factor (TNF) decreased upon dosing of anti-TNF antibody drugs such as adalimumab, infliximab, certolizumab, and etanercept. However, quantification of free cytokines is challenging due to low sensitivity and high variability. Under such circumstances, it was recommended to measure total (unbound plus bound) concentrations and use the PK/PD model to calculate free levels. ELISA methods are designed to analyze the free or total ligand and the method specificity should be properly confirmed on the measured forms [8]. The effect of target binding to an LM drug can be characterized with both free and total measurements to build preclinical PK/PD models [9].

(B) Membrane receptor sheds soluble fragments into the circulation—The extracellular domain of a transmembrane receptor clipped by proteolysis is often found in blood (e.g., bacterial or cancer cell markers). If the target is a cell membrane receptor located in cells or organs that are not clinically accessible, the indirect measurement of the shed fragments in blood can be made. The animal model provides an opportunity to obtain direct evidence of the target hit and engagement of the drug at the cell membrane receptors in tissues and correlate to the indirect measurements of the shed fragments in blood [10].

(C) Target inside cells—These cells would not be easily obtained in a clinical study and the animal model offers a relevant system for investigation. Cells or tissues should be carefully excised from the animals. If the membrane target also exists in accessible blood or skin cells, these cells can be used as a surrogate matrix for measurement. The target can be extracted and released from the cells for assay. Care must be taken not to create artifacts from stimulation or suppression over the course of specimen collection, storage, shipment, extraction, and analysis. Considerations are similar to those discussed for biomarkers by Lee et al. [11].

The in vivo binding of the drug to the target ligand is characterized by the percent target occupancy, which is a parameter besides the dissociation equilibrium constant K_d in the PK/PD model. Receptor (target) occupancy is the first component of functional target engagement for proof of mechanism, followed by subsequent modulations of downstream biomarkers [12]. The in vitro K_d is measured with binding tests using radiolabeled ligand or surface plasmon resonance (BiaCore). The in vivo K_d is usually estimated from PK exposure and target ligand concentrations. Target occupancy can be measured with positron emission tomography (PET) with radiolabeled ligands. PET scanners have been designed specifically for imaging rodents and small primates [13]. The technique has been used for SM candidates. However, the application of radiolabel techniques to LM candidates is challenging due to difficulties of making the labeled protein drug with the same binding activity.

Biomarkers that show target hit and subsequent responses related to dosing from preclinical studies are selected for further clinical investigation. Several complex human diseases with multiple causes have used animal models for evaluation. For the development of cancer drug target on tyrosine kinase inhibition, kinase phosphorylation is a target engagement biomarker. While survival is the endpoint in a preclinical study, additional parameters such as metabolic rate and tumor volume/size are often evaluated [14]. Another example is the cardiovascular disorder caused by dysfunctional cholesterol metabolism. The recent drug development of proprotein convertase subtilisin/kexin type 9 (PCSK9) has successfully applied mouse models of deletion or different levels of PCSK9 expression to show that PCSK9 binding to LDL receptors led to receptor degradation, causing the decrease in LDL cholesterol clearance [15]. Plasma PCSK9 can be used as the target hit biomarker and the LDL cholesterol as the downstream biomarker for drug candidate screening and then for dosing design [16].

Pathway Consequences

Biomarkers that are associated with the target are "proximal biomarkers"; those that are further downstream are "distal biomarkers." Figure 1 depicts a simplified pathway

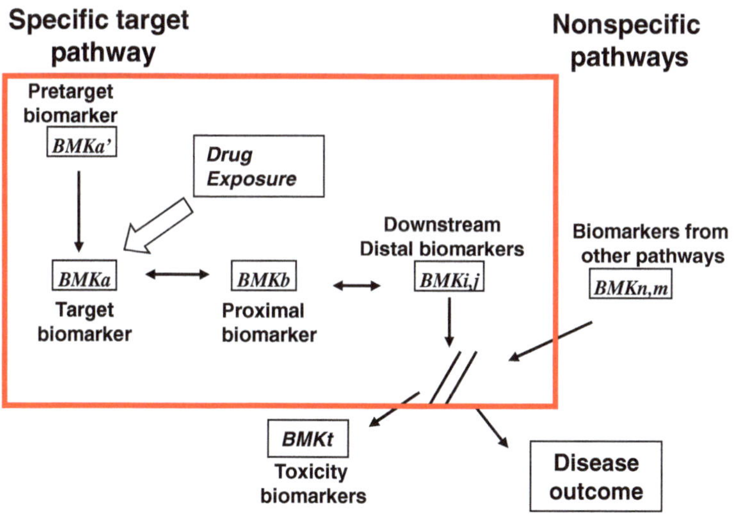

Fig. 1 Scheme of biomarker pathways. The target-related pathways and drug exposure effects are shown inside the red square; the nonspecific pathways are shown outside

scheme. The target-related pathways and drug exposure effects are shown inside the red square: The target biomarker BMKa may be the product of proteolysis from a precursor protein BMKa'. Drug interaction with the target may affect a proximal biomarker BMKb. Further downstream distal biomarkers BMKi,j are modulated and specifically affect the disease outcome. The nonspecific pathways are shown outside the red square: Other pathways of different mechanism of action may affect the disease outcome with changes in biomarkers $BMK_{n,m}$. In addition, on- or off-target actions can illicit toxic effects that can be monitored by toxicity biomarkers BMKt.

Animal species may not have exactly the same biochemical pathways as that of humans. Even among humans, intrinsic and extrinsic variations can lead to different pathway consequences and biomarker responses. Focus on the drug target pathway to select biological relevant animal model with appropriate PD biomarkers is essential for PK/PD modeling. Multiple proximal and distal biomarkers offer choices of PD biomarkers to test the pathway hypothesis and track the consequences. During preclinical studies, a panel of biomarkers is often monitored. Those with a dose-response relationship are selected for further use. For example, during the development of denosumab, that target is the nuclear factor-κB ligand (RANKL), the proximal biomarker is a decoy receptor osteoprotegrin (OPG), and distal biomarkers are bone specific alkaline phosphatase (BSAP), N-terminal propeptide of type 1 procollagen (P1NP), tartrate-resistant acid phosphatase 5b (TRACP-5b), intact parathyroid hormone (iPTH), and serum C-telopeptide (sCTx) and *N*-telopeptide of type I collagen (NTx). The latter two biomarkers were found to be suppressed in a dose dependent manner in rodents and validated methods were further used in clinical studies as PD biomarkers [17, 18]. In addition, multiple sets of biomarkers may be statistically stronger than that of a single set in decision-making and confirming the pathway hypothesis. The PD data can be used to estimate effective dose range differentiating from those of toxic side effects for drug candidate selection and lessons learned from structural effects.

The necessity to analyze multiple biomarkers drives the development of multiplex methods and instruments such as Luminex and Mesocale Discovery (MSD) [19]. For biomarkers of the animal species that do not share sufficient homology with humans, the production of reagents for immunoassays would require enormous time and resource. Mass spectrometric (MS) methods have recently emerged for protein biomarker quantification [20]. The protein analyte is enzymatically digested to peptides. With known specific peptide sequences of the animal species, a unique peptide can be selected to be quantified by high performance liquid chromatography (LC) MS. The method does not require specific binding reagents while it offers an advantage of structural specificity. For preclinical studies involving disease models of several animal species, the homolog peptide of a different species can be quantified similarly with minimal method development time. After the few biomarkers have been chosen, resources can be directed to reagents procurement and ELISA method development.

Determinations in Circulatory and Tissue Spaces

Biomarkers are commonly chosen based on their accessibility, such as those in circulation. For diseases that are localized at internal tissues, two questions are often asked: 1. Will there be sufficient drug levels in the disease tissue to sustain the effect? 2. Can the drug effect be measured at the tissue? The first question is answered by target occupancy at the tissue. For the second question, indirect data of soluble biomarkers in circulation are often used to infer the drug effect on tissue target. Animal studies offer an opportunity to confirm this inference. Blood and tissue samples can be collected over time in a preclinical study to determine drug and biomarker concentrations. PK/PD parameters, including target occupancy, can be calculated from blood and tissue concentrations to build mechanistic model such as the physiological-based PK (PBPK) model [21].

Tissue imaging contributes to the understanding of the relationship of drug structures to target engagement effect in the tissue space. Quantitative radioautography has been used for whole body distribution of SM drug compounds in animal models. Recent development of sensitive and high resolution molecular imaging with fluorescent probes has been developed [22]. These probes can be designed to retain bioactivities with sufficient sensitivity for time-lapse living cells imaging to add PD information at the organ level [23]. Emerging MS methods do not require labeling. The use of matrix assist laser desorption ionization (MALDI) MS can provide information on structural and spatial changes in protein biomarkers expressions together with drug distribution [24].

Safety Biomarkers—From General Physiological Markers to Mechanistic-Based Safety Biomarkers

The Critical Path Initiative of the US Food and Drug Administration has propelled the development of both efficacy and safety biomarkers. Similar to the function of efficacy biomarkers, safety biomarkers inform on the mechanism of toxicity and early detection and prediction of adverse effects. For example, the Critical Path Institute's Predictive Safety Testing Consortium (PSTC) have concluded seven urinary proteins (KIM-1, albumin, total protein, β_2-microglobulin, cystatin C, clusterin, trefoil factor-3) to be considered as qualified biomarkers of kidney injury in regulatory decision-making. The kidney biomarkers are acceptable in the context of nonclinical drug development for the detection of acute drug-induced kidney toxicity. Other working groups are making progress on the development of safety biomarkers on cardiac hypertrophy, liver, skeletal muscle, testicular toxicity, and vascular injury [25].

The PK/PD, adverse event, and histopathological data in preclinical studies enable the identification of safety biomarkers that precede the manifestation of histological lesions and prediction of severity. To qualify them as predictive safety biomarkers, time course and dose-response studies across species must be carried out with robust positive and negative controls for preclinical evaluation. In addition, comparable biological responses between nonclinical species and humans should be established with considerations in the limitations of specimen collections in clinical trials and method/technology translation.

Integral Informative Assessment of Risks Versus Benefits to Influence Decisions

Drug development in the pharmaceutical industry requires teamwork. A drug program development team may consist of pharmacologists, toxicologists, clinicians, PK/PD and bioanalytical scientists. Biomarker and drug exposure data are generated from the bioanalytical laboratories and primarily used by the PK/PD scientists. During the early phase of development, the team often agrees on a biomarker plan with potential efficacy and safety biomarkers to test risk and benefit to inform decision-making. At this time, limitations in reagents and time pressure would be considered for the biomarker list and method approach. The bioanalytical scientists should understand the intended purposes of the biomarkers to develop and validate fit-for-purpose assays and clearly communicate the assay characteristics to the stake holders to assure correct applications [11].

In general, large data sets are accumulated on multiple candidates from various series of chemical structures, their metabolic liabilities, multiple efficacy and safety biomarkers, as well as qualitative and quasi-quantitative data of histology and physiological measurements. Multivariate statistical techniques, with correlations for multiple tests, can be used to analyze large data sets. In addition, bioinformatics techniques such as principal component analysis and hierarchical cluster analysis can be used to detect relationships among multiple analytes and specimens [26, 27]. These topics are out of the scope of this chapter.

Animal studies provide an opportunity to collect data of circulatory concentrations and also those of various organs in the test system. The data are useful in building PBPK models that have been recommended over traditional TK/PK/PD models for better prediction and translation from animal to human [21]. Several PBPK models were proposed with considerations of organ size, circulatory flow, degradation, and intracellular pH equilibrium and transit times for monoclonal antibody drugs [28, 29]. The incorporation of the PD component in the model requires complex consideration of the biological pathways, including the dynamics of target synthesis and elimination, cell cycles, and physiological responses [30, 31]. The PBPK models are also recommended for the prediction of drug–drug interactions to be tested in the relevant animal species [32].

Summary

- Overview: Biomarkers data are necessary for preclinical toxicokinetics/pharmacokinetics/pharmacodynamics model building during drug development for drug candidate selection and to aid the first time in human starting dose decision and clinical study design.
- Animal species with biological relevancy to humans are chosen for preclinical studies. Biomarkers reflecting various biological pathways are used to confirm the relevancy and test the drug effect hypothesis.
- Efficacy is demonstrated by changes on the target biomarkers and pathway consequences that are related to drug concentrations in circulation and tissues.
- Preclinical studies enable the identification and application of safety biomarkers to predict adverse effect.
- Physiological-based pharmacokinetics/pharmacodynamics modeling has been recommended for the integral assessment of risks versus benefits to influence decisions.

References

1. Berg JM, Rogers ME, Lyster PM (2010) Systems biology and pharmacology. Clin Pharmacol Ther 88:17–19
2. FDA Mar 2004 report. Innovation or Stagnation? Challenge and opportunity on the critical path to new medical products. http://c-path.org/pstc.cfm
3. Piccotti JR, Lebrec HN, Evans E et al (2009) Summary of a workshop on nonclinical and clinical immunotoxicity assessment of immunomodulatory drugs. J Immunotoxicol 6:1–10
4. Suntharalingam G, Perry MR, Ward S et al (2006) Cytokine storm in a phase 1 trial of the anti-CD28 monoclonal antibody TGN1412. N Engl J Med 355:1018–1028
5. Richmond A, Su Y (2008) Mouse xenograft models vs. GEM models for human cancer therapeutics. Dis Model Mech 1:78–82
6. Chatzigeorgiou A, Halapas A, Kalafatakis K et al (2009) The use of animal models in the study of diabetes mellitus. In Vivo 23:245–258
7. King AJ (2012) The use of animal models in diabetes research. Br J Pharmacol 166:877–894
8. Lee JW, Kelley M, King LE et al (2011) Bioanalytical approaches to quantify “total” and “free” therapeutic antibodies and their targets: technical challenges and PK/PD applications over the course of drug development. AAPS J 13:99–110
9. Betts AM, Clark TH, Yang J et al (2010) The application of target information and preclinical pharmacokinetic/pharmacodynamic modeling in predicting clinical doses of a dickkopf-1 antibody for osteoporosis. J Pharmacol Exper Therap 333:2–13
10. Park NJ, Wang X, Diaz A et al (2013) Measurement of cetuximab and panitumumab-unbound serum EGFR extracellular domain using an assay based on slow off-rate modified aptamer (SOMAmer) reagents. PLOS ONE 8(8):e71703–e71703
11. Lee JW, Devanarayan V, Barrett YC et al (2006) Fit-for-purpose method development and validation for successful biomarker measurement. Pharm Res 23:312–328
12. Morgan P, Van Der Graaf PH, Arrowsmith J et al (2012) Can the flow of medicines be improved? Fundamental pharmacokinetic and pharmacological principles toward improving Phase II survival. Drug Discov Today 17:419–424

13. Zhang Y, Fox GB (2012) PET imaging for receptor occupancy: meditations on calculation and simplification. J Biomed Res 26(2):69–76
14. Visser EP, Philippens ME, Kienhorst L et al (2008) Comparison of tumor volumes derived from glucose metabolic rate maps and SUV maps in dynamic 18F-FDG PET. J Nucl Med 49 (6):892–898
15. Maxwell KN, Breslow JL (2004) Adenoviral-mediated expression of Pcsk9 in mice results in a low-density lipoprotein receptor knockout phenotype. Proc Natl Acad Sci USA 1017100–7105
16. Graham MJ, Lemonidis KM, Whipple CP et al (2007) Antisense inhibition of proprotein convertase subtilisin/kexin type 9 reduces serum LDL in hyperlipidemic mice. J Lipid Res 48:763–767
17. Silva-Fernándeza L, Rosariob MP, Martínez-Lópezc JA et al (2013) Denosumab for the treatment of osteoporosis: a systematic literature review. Reumatol Clin 9:42–52
18. Wang J, Lee JW, Burns D et al (2009) "Fit-for-Purpose" method validation and application of a biomarker (C-terminal telopeptides of type-1 collagen) in Denosumab clinical studies. AAPS J 11:385–394
19. Chowdhury F, Williams A, Johnson P (2009) Validation and comparison of two multiplex technologies, Luminex and Mesoscale Discovery, for human cytokine profiling. J Immunol Methods 340:55–64
20. Ezan E, Dubois M, Becher F (2009) Bioanalysis of recombinant proteins and antibodies by mass spectrometry. Analyst 134:825–834
21. Rowland M, Peck C, Tucker G (2011) Physiologically-based pharmacokinetics in drug development and regulatory science. Annu Rev Pharmacol Toxicol 51:45–73
22. Hoppin J, Orcutt KD, Hesterman JY et al (2011) Assessing antibody pharmacokinetics in mice with in vivo imaging. J Pharmacol Exp Ther 337:350–358
23. Close DM, Xu T, Sayler GS et al (2011) In vivo bioluminescent imaging (BLI): noninvasive visualization and interrogation of biological processes in living animals. Sensors 11:180–206
24. Wang KY, Chuang SA, Lin PC et al (2008) Multiplexed immunoassay: quantitation and profiling of serum biomarkers using magnetic nanoprobes and MALDI-TOF MS. Anal Chem 80:6159–6167
25. Quantitative and Systems Pharmacology Workshop II: http://meetings.nigms.nih.gov/index.cfm?event=home&ID=8316
26. Wong HL, Pfeiffer RM, Fears TR et al (2008) Reproducibility and correlations of multiplex cytokine levels in asymptomatic persons. Cancer Epidemiol Biomark Prev 17:3450–3456
27. Khan IH, Krishnan VV, Ziman M et al (2009) A comparison of multiplex suspension array large-panel kits for profiling cytokines and chemokines in rheumatoid arthritis patients. Cytom B Clin Cytom 76:159–168
28. Shah DK, Betts AM (2012) Towards a platform PBPK model to characterize the plasma and tissue disposition of monoclonal antibodies in preclinical species and human. J Pharmacokinet Pharmacodyn 3967–3986
29. Chen Y, Balthasar JP (2012) Evaluation of a catenary PBPK model for predicting the in vivo disposition of mAbs engineered for high-affinity binding to FcRn. AAPS J 14:850–859
30. Thygesen P, Macheras P, Van Peer A (2009) Physiologically-based PK/PD modelling of therapeutic macromolecules. Pharm Res 26:2543–2550
31. Krzyzanski W, Sutjandra L, Perez-Ruixo JJ et al (2013) Pharmacokinetic and pharmacodynamic modeling of romiplostim in animals. Pharm Res 30:655–669
32. Huang SM, Zhao H, Lee JI et al (2010) Therapeutic protein-drug interactions and implications for drug development. Clin Pharmacol Ther 87:497–503

Incorporating Clinical Biomarkers into Clinical Trials

Paul W. Rhyne

Abstract Clinical trials are required for testing the safety and efficacy of new drugs produced by the pharmaceutical industry. These trials also serve to provide information on clinical and survival benefits, prediction of treatment responses, identification of patient subpopulations that will benefit from the drug, and many other important aspects of treatment. Clinical biomarkers are essential tools in these trials that enable these different aspects to be evaluated and defined. However, the incorporation of biomarkers into clinical trials requires a knowledge base that includes understanding the different types of clinical biomarkers, selection of the best biomarkers for the trial, best sample handling and processing practices for the biomarker, validation planning, selection of the best technology platform, and selection of the best laboratory to perform the analysis. This chapter provides an overview of how clinical biomarkers are used in clinical trials and discusses the different aspects of incorporating them into clinical trials.

Keywords

Biomarker A characteristic that is objectively measured and evaluated as an indicator of normal biologic processes, pathogenic processes, or pharmacologic responses to a therapeutic intervention

Clinical trial Research-based studies involving human volunteers that are assigned to receive one or more interventions so that researchers can evaluate the effects of the interventions on biomedical or health-related outcomes

Validation Confirmation through laboratory testing that the performance characteristics of an assay are suitable and reliable for its intended analytical use

Critical reagent Reagents such as antibodies, oligonucleotides, enzymes, or fluorescent molecules that are integral parts of an assay that influence assay performance or quality

P.W. Rhyne (✉)
Immunoassay Services, Q[2] Solutions, 1600 Terrell Mill Road, Suite 100, Marietta, GA 30067, USA
e-mail: paul.rhyne@q2labsolutions.com; paul.rhyne@yahoo.com

R. Weiner and M. Kelley (eds.), *Translating Molecular Biomarkers into Clinical Assays*, AAPS Advances in the Pharmaceutical Sciences Series 21,
DOI 10.1007/978-3-319-40793-7_6

Why Incorporate Biomarkers into Clinical Trials?

A question that is often asked by clinical teams is why should biomarkers be incorporated into clinical trials? The answer to this question starts by defining exactly what a clinical trial is. The US National Institutes of Health defines clinical trials as research-based studies involving human volunteers that are "assigned to receive one or more interventions so that researchers can evaluate the effects of the interventions on biomedical or health-related outcomes". In these studies, participants receive specific interventions according to the research plan or protocol created by the investigators [18]. Clinical trials may include testing new medical products, new drugs, new treatment procedures, or comparing new medical approaches to existing ones. A major aspect of clinical trials is the evaluation of safety and efficacy of these interventions in the participants. However, there are other aspects of clinical trials equally important such as prediction of treatment benefit, evaluation of survival benefit, selection of drug dosing, demonstrating clinical benefit, verification of the therapeutic biological target, identification of patient subpopulations, etc. Incorporating biomarkers into clinical trials provides the necessary tools to evaluate many of these other important aspects in addition to drug safety and efficacy.

Biomarkers

Given the increased use of biomarkers in clinical trials, it is important to understand how biomarkers are defined and the differences between the different categories. The term biomarker has been defined as a characteristic that is objectively measured and evaluated as an indicator of normal biologic processes, pathogenic processes, or pharmacologic responses to a therapeutic intervention [1]. The National Cancer Institute Investigational Drug screening task force created a biomarker task force that was to provide recommendations for the use of biomarkers in clinical trials [2]. The task force also provided some definitions on different subcategories of biomarkers. Prognostic biomarkers provide evidence about the patients overall disease outcome independent of any specific intervention. Predictive biomarkers provide evidence about the probability of benefit or toxicity from a specific intervention. Surrogate biomarkers are intended to serve as a substitute for a clinically meaningful endpoint. Pharmacodynamic biomarkers are used to provide evidence of pharmacological effects of drugs [2]. Diagnostic biomarkers provide information that aids to establish or confirm a diagnosis. Exploratory biomarkers are used mainly for hypothesis generation. They are typically based on scientific literature and knowledge of biological pathways, and have not previously been shown to have clinical significance. There are so many different subcategories of biomarkers that have been described, it is important to remember that they are not always referring to the same thing.

Predictive Biomarkers

The incorporation of biomarkers into clinical trials can provide tools to predict how individual patients will respond or benefit from treatment. Clinical trials that incorporate the use of predictive biomarkers often involve pretesting potential trial subjects prior to enrollment. In some cases, only those subjects that have a "positive" or "negative" biomarker profile will be enrolled into the trial. This may be used to eliminate subpopulations that will not benefit from treatment. Alternatively, there may be a concern for putting a potentially harmful therapeutic drug into a non-disease state population. This is often the case for oncology and infectious diseases therapeutics, where the treatments create a safety concern in healthy individuals. Predictive biomarkers are typically genetic-based biomarkers such as KRAS [3] or BRAF [4]. However, other types of predictive biomarkers are increasingly being used such as cellular phenotyping [5], expression levels of proteins [6, 7], and others. Predictive biomarkers are also used during the trial to assess progress and subsequently to change aspects of the trial midstream. Examples of these type trial designs include BATTLE (biomarker integrated approaches of targeted therapy for non-small cell lung Carcinoma) [8] and I-SPY [9]. Trials can also be designed with the purpose of codeveloping a predictive biomarker with the therapeutic drug being developed. These require a robust hypothesis and high-quality clinical data, so that benefit and magnitude of the benefit of treatment can be determined.

Selection of Clinical Biomarkers

The development of new drug products involves a significant amount of investment in drug discovery and preclinical work, before clinical trials can be initiated. Biomarkers are incredibly valuable in these early stages and are often used to help evaluate early drug candidates before clinical trials begin. In the early stages of clinical trial planning, clinical teams need to identify biomarkers that will be used in these trials. Selection of potential clinical biomarkers include those identified cell lines, preclinical animal studies, those used in other clinical studies, and the published literature. Consideration should be given to ensure the selected biomarker is amenable for clinical use. Laboratory scientists performing these studies have control over the cellular growth conditions including viability, drug exposure, and cell numbers in each of the experiments. In contrast, this level of control is not available for samples collected from clinical trial participants. The expression level of the biomarker in clinical samples needs to be evaluated to ensure the biomarker is present in sufficient quantities to be useful.

Preclinical studies in cell lines often involve stimulation or treatment of cells to induce changes in the biomarker [10]. Performing these treatments or providing stimulation to cells from cell lines are relatively easy in comparison to using isolated cells taken from clinical trial participants. The clinical sites where trial participants are being treated often do not have the needed expertise or the right equipment to implement complex pre-analytical sample collection and handling procedures. Isolating cells from blood and stimulating them at the clinical sites are often met with significant technical challenges. Another option is to send the samples to an external lab with the required expertise in the methodology. This will require additional logistics for the shipment of the samples to a different location which may expose the samples to conditions that change the cellular response generated from treatment [11, 12]. Finally, cellular responses to the stimulation conditions used on cell lines are likely to be different then in freshly isolated cells [10]. Thus, it is recommended that some level of investigation be done on cell-based biomarkers from freshly isolated cells before incorporating them into the clinical trial.

Clinical Biomarker Pre-analytic Sample Collection and Handling

One of the key attributes of incorporating clinical biomarkers into clinical trials is the collection and processing of the clinical samples. Biomarker assays are typically developed, tested, and validated using common laboratory buffers which do not reflect the clinical samples. The assay should be tested in the same matrix as the clinical samples before being used for sample analysis. The sample collection procedures, processing, and handling of the clinical samples has an impact on the biomarker measurement [12]. A prime example would be VEGF. While VEGF has measureable expression levels in serum and plasma, care must be taken in the collection and processing of the clinical samples to avoid inadvertent release of VEGF from platelets [11]. Improper sample collection will result in levels of VEGF that do not reflect the status of the test subject, but rather improper handling at the clinical site. Another example would be the addition of compounds or reagents directly to the clinical sample [13]. These additives are often added to a specified volume of the clinical sample, in which the premeasured volume may not be accurate. Thus, careful evaluation of the instrumentation and staff expertise at the clinical site should be done to ensure successful handing and processing of samples. This evaluation should be repeated at regular intervals, every 6–12 months and/or when key site personnel change.

Biomarker Assay Performance and Prevalidation Planning

Biomarker assays should have good clinical utility and have performance characteristics that are suitable for clinical trials. The first step in selecting the best assay is to define what question the biomarker data will be used to answer. This will be used to define the required performance characteristics of the assay including analytical sensitivity, accuracy and precision, range of detection, and dilutional linearity [2, 14]. Once the performance requirements are defined, the search for an appropriate assay can begin. The assay format, technology base, availability of assay reagents, cost, level of technical expertise, lot-to-lot variability of kits and reagents, and data format should be considered in the selection of the best assay. The clinical team should strive for simple assay formats in lieu of more complex platforms, which are more difficult to perform and troubleshoot. The technology used in the assay will determine instrumentation needed, which could limit the number of available labs capable of running the assay. Technologies that are well established tend to be more widely available in laboratories than newer technologies. Newer technologies may offer or claim advantages over existing technologies, but often come with increased operating costs and the need for additional technical expertise technical expertise. Newer technologies also carry an increased risk of unknown or unresolved technical issues that may be encountered during the clinical trial. Thus, it is essential to have access to expertise to minimize this impact. Utilizing a technology that is provided by a single company carries an economic risk and a risk of not being able to complete sample analysis. There are examples of single source vendors that have gone out of business or are no longer able to support their technology. Cost of the instrumentation and needed reagents/supplies should be assessed before final selection of the technology. This is especially critical for highly labile samples where storing the sample for an extended period of time, while an alternative method is identified, is not feasible due to lack of long-term sample stability.

Importance of Critical Reagents

Biomarker assays, regardless of the assay format or the technology, rely on critical analytical reagents such as antibodies, oligonucleotides, enzymes, or fluorescent molecules. It is essential that these critical reagents are available throughout the clinical trial period in which the biomarker assay is used. Evaluation of supply and expiration dates should be done well before biomarker analysis is started [15, 16]. Protein-based reagents are often given an expiration date of one year, which presents problems for clinical trials that extend beyond a year. As a result, this would require multiple lots of reagents to be purchased and used. It is incumbent upon the bioanalytical scientist to evaluate multiple lots of reagents and determine if all lots perform as the original lot of reagent. In addition to reagents, other critical supplies such as assay plates, chips, and disposables fall into this category. Careful planning will ensure that delays from back ordered items are minimized as much as possible.

Lot changes of critical reagents and supplies often impact the performance of the assay, which can result in significant differences in reported biomarker results.

Examples of this include differences in precoated plates such as streptavidin plates, changes in lot of antibody, differences in the labeling of antibodies used to detect the biomarker, or differences in purity of critical reagents such as peptides or oligonucleotides. Instances where lot changes occur should be evaluated in bridging experiments to assess the impact on biomarker assay performance and reported results. This can be disruptive during a clinical trial that is dependent on biomarker data for enrollment. Careful planning is needed to minimize the impact of availability and lot changes during the course of the trail.

Biomarker Data Handling

Analyzing biomarkers in clinical samples will generate data that is reported back to the clinical team. Some level of evaluation must take place before sample analysis begins. The handling and generation of clinical data must be compliant with good clinical laboratory practices (GCP) [19], which are typically handled through specialized software that protects patient identification, reported results, and other important information in a secure manner. The transfer of biomarker data into clinical databases is not trivial and can be the rate-limited step to getting data to the clinical team. Data may need to be transformed into various formats such as text or comma separated values (csv) before the data can be imported into clinical databases. The final data should be checked for transcription errors that may have been occurred during the transfer of data into the clinical database.

Commercial Biomarker Assays

The selection of a suitable biomarker assay for clinical sample analysis is sometimes easier when the assay is available as a commercial kit. The ability to purchase a premade kit eliminates the need for assay development and simplifies the process of reagent procurement. Commercially available kits provide prewritten assay protocols, prepackaged reagents and supplies, and technical support from the kit manufacturer all with a fixed cost. Commercial kits that have the required assay performance needs provide an excellent and viable option for the clinical team. However, there are several risks that come with selecting a commercial kit for biomarker analysis [17]. Manufacturers of commercial kits often develop the kits using laboratory buffers and not clinical matrices such as serum or plasma. These kits require testing in the appropriate biological matrix before they can be reliably used to support clinic sample analysis. There are kit manufacturers that recognize this and have started to provide kit performance data in biological matrices. Clinical teams selecting commercial sources of biomarker assays should make sure the assay can measure the biomarker in samples that are as close as possible to the clinical samples taken from trial participants.

Biomarker Assay Validation

Once a suitable biomarker assay has been developed or acquired, the clinical team should have the assay validated prior to use. The required performance characteristics should be used to generate a validation plan. The amount of effort, time, and performance characteristics tested for each biomarker assay is often referred to as a "fit-for-purpose" validation plan [14]. In some cases where the biomarker data is for informational or exploratory in nature, a minimal amount of validation may be appropriate. In other cases, where the biomarker will be used as clinical surrogate endpoints, used for enrollment or dosing decisions, or go/no-go decisions the level of analytical validation should be advanced and more comprehensive. The data generated during validation should be evaluated to ensure the assay meets the performance requirements and provide useful data to the clinical team before the assay is used in the clinical trial. The assay should also have run acceptance criteria that are used throughout sample analysis to ensure the assay performance is consistent. The performance of the assay throughout the sample analysis period should be reviewed and used to define the clinical performance characteristics of the assay for subsequent and future clinical trial use.

Selection of the Biomarker Laboratory

Selecting the right biomarkers and the best assays for a clinical trial should be done with careful planning and evaluation. However, selecting the best laboratory to perform the work is equally important. Several aspects of laboratory selection that need to be part of the decision process include instrumentation, technical expertise, SOPs, certifications, and cost. The majority of biomarker laboratories are equipped with instrumentation for well-established technologies, but may not always have newer technologies. Clinical teams should perform some level of laboratory qualification to ensure the laboratory has the right instrumentation, maintains the instrumentation in good working order, and keeps good maintenance records for the instrumentation. The qualification process should also include a review of laboratory staff qualifications (training, education, and experience) to ensure the level of expertise is present to run the assay correctly. Laboratories with experienced staff are likely to solve technical issues more efficiently and identify potential problems with the assay before they occur. Biomarker laboratories should have documented SOPs and work practices in place to ensure consistency in running assays regardless of the scientist performing the work.

Laboratory Certification and Operational Practices

There are multiple types of certifications and practices that laboratories can acquire or implement. Incorporating biomarkers into clinical trials may or may not require laboratory certification. Thus, clinical teams should have a working knowledge of these certifications and practices in addition to the requirements for biomarker analysis. A few examples of these would be Good Clinical Practice (GCP) [19], Good Laboratory Practice (GLP) [20], and certifications under the Clinical Laboratory Improvement Amendments (CLIA) [21].

GLP is for laboratories conducting nonclinical laboratory studies, which do not involve human subjects [20]. GLP ensures that all laboratory testing is performed by qualified personnel in adequate facilities and supervision. The equipment has to be well maintained, 21 CSF Part 11 compliant and calibrated prior to use. Written SOPs must be in place and all work is fully documented to ensure traceability and reproducibility. GLP also requires monitoring of the study by a separate, quality assurance unit. Good Clinical Practice, or GCP, provides a standard for the design, conduct, performance, monitoring, auditing, recording, analysis, and reporting of clinical trials. GCP provides assurance that the data and reported results are credible and accurate. Also, GCP provides assurance that the rights, integrity, and confidentiality of the trial subjects are protected. In 1988, congress passed the clinical laboratory improvement amendments (CLIA) [21]. CLIA establishes quality standards for all non-research laboratory testing performed on specimens derived from humans for the purpose of providing information for the diagnosis, prevention, treatment of disease, or impairment of, or assessment of health. Thus, if a biomarker assay is to be used on human samples for the diagnosis, prevention, or treatment of disease, the laboratory performing the assay would need to be CLIA certified and hold a CLIA certificate that corresponds to the test being performed. Laboratories that have a CLIA certification should have well-maintained instrumentation, documented training of staff, background information on staff, and have good documentation on reagents. Depending on the intended use of the biomarker, the lab may be required to implement GLP or GCP practices, or obtain CLIA certification.

Clinical trials are essential tools in the drug development and approval process. A significant amount of effort is needed to ensure a successful clinical trial outcome. Biomarkers can be used to predict treatment outcome, adjust drug dosing during the trial, verify targeted biological pathways, provide information of drug safety and set subject enrollment criteria, all of which increase the success rate of the clinical trial. Thus, the incorporation of biomarkers in clinical trials plays a crucial part in the success rate of clinical trials and subsequently the drug development process.

Chapter Summary

1. There are many types and definitions of clinical biomarkers.
2. Sample collection and handling should be carefully planned to minimize the impact on the measurement of the biomarker.
3. Assays used to measured clinical biomarkers should be evaluated with a fit-for-purpose validation plan to ensure the assay is suitable for the clinical trial.
4. Biomarker laboratories should have the proper instrumentation, experience, and work practices that are needed to support the clinical trial.

References

1. Atkinson AJ, Colburn WA, DeGruttola VG, DeMets DL, Downing GJ, Hoth DF, Oates JA, Peck CC, Schooley RT, Spilker BA, Woodcock J, Zeger SL (2001) Biomarkers and surrogate endpoints: preferred definitions and conceptual framework. Clin Pharmacol Ther 69(3):89–95
2. Dancey JE, Dobbin KK, Groshen S, Jessup JM, Hruszkewycz AH, Koehler M, Parchment R, Ratain MJ, Shankar LK, Stadler WM, True LD, Gravell A, Grever MR (2010) Guidelines for the development and incorporation of biomarker studies in early clinical trials of novel agents. Clin Cancer Res 16:1745–1755
3. Lièvre A, Bachet JB, Le Corre D, Boige V, Landi B, Emile JF, Côté JF, Tomasic G, Penna C, Ducreux M, Rougier P, Penault-Llorca F, Laurent-Puig P (2006) KRAS mutation status is predictive of response to cetuximab therapy in colorectal cancer. Cancer Res 66:3992–3995
4. Davies H, Bignell GR, Cox C, Stephens P, Edkins S, Clegg S, Teague J, Woffendin H, Garnett MJ, Bottomley W, Davis N, Dicks E, Ewing R, Floyd Y, Gray K, Hall S, Hawes R, Hughes J, Kosmidou V, Menzies A, Mould C, Parker A, Stevens C, Watt S, Hooper S, Wilson R, Jayatilake H, Gusterson BA, Cooper C, Shipley J, Hargrave D, Pritchard-Jones K, Maitland N, Chenevix-Trench G, Riggins GJ, Bigner DD, Palmieri G, Cossu A, Flanagan A, Nicholson A, Ho JW, Leung SY, Yuen ST, Weber BL, Seigler HF, Darrow TL, Paterson H, Marais R, Marshall CJ, Wooster R, Stratton MR, Futreal PA (2002) Mutations in the BRAF gene in human cancer. Nature 417:949–954
5. Younes A, Gopal AK, Smith SE, Ansell SM, Rosenblatt JD, Savage KJ, Ramchandren R, Bartlett NL, Cheson BD, de Vos S, Forero-Torres A, Moskowitz CH, Connors JM, Engert A, Larsen EK, Kennedy DA, Sievers EL, Chen R (2012) Results of a pivotal phase II study of brentuximab vedotin for patients with relapsed or refractory Hodgkin's lymphoma. J Clin Oncol 30(18):2183–2190
6. Lanari A, Parnetti L (2009) Cerebrospinal fluid biomarkers and prediction of conversion in patients with mile cognitive impairment: 4-year follow-up in a routine clinical setting. Sci World J 9:961–966
7. Mattsson N, Zetterberg H, Hansson O, Andreasen N, Parnetti L, Jonsson M, Herukka SK, van der Flier WM, Blankenstein MA, Ewers M, Rich K, Kaiser E, Verbeek M, Tsolaki M, Mulugeta E, Rosén E, Aarsland D, Visser PJ, Schröder J, Marcusson J, de Leon M, Hampel H, Scheltens P, Pirttilä T, Wallin A, Jönhagen ME, Minthon L, Winblad B, Blennow K (2009) CSF biomarkers and incipient Alzheimer disease in patients with mild cognitive impairment. J Am Med Assoc 302:385–393

8. Kim ES, Herbst RS, Wistuba II, Lee JJ, Blumenschein GR Jr, Tsao A, Stewart DJ, Hicks ME, Erasmus J Jr, Gupta S, Alden CM, Liu S, Tang X, Khuri FR, Tran HT, Johnson BE, Heymach JV, Mao L, Fossella F, Kies MS, Papadimitrakopoulou V, Davis SE, Lippman SM, Hong WK (2011) The BATTLE trial: personalizing therapy for lung cancer. Cancer Discovery 1:44–53 (2011)
9. Esserman LJ, Berry DA, Cheang MC, Yau C, Perou CM, Carey L, DeMichele A, Gray JW, Conway-Dorsey K, Lenburg ME, Buxton MB, Davis SE, van't Veer LJ, Hudis C, Chin K, Wolf D, Krontiras H, Montgomery L, Tripathy D, Lehman C, Liu MC, Olopade OI, Rugo HS, Carpenter JT, Livasy C, Dressler L, Chhieng D, Singh B, Mies C, Rabban J, Chen YY, Giri D, Au A, N Hylton (2012) Chemotherapy response and recurrence-free survival in neoadjuvant breast cancer depends on biomarker profiles: results from the I-SPY 1 TRIAL (CALGB 150007/150012; ACRIN 6657). Breast Cancer Res Treat 132:1049–4062
10. Schildberger A, Rossmanith E, Eichhorn T, Strassl K, Weber V (2013) Monocytes, peripheral blood mononuclear cells, and THP-1 cells exhibit different cytokine expression patterns following stimulation with lipopolysaccharide. Mediators Inflamm 2013:1–10
11. Schlingemann RO, Van Noorden JF, Diekman MJM, Tiller A, Meijers JCM, Kookwijk P, Wiersinga WM (2013) VEGF levels in plasma in relation to platelet activation, glycemic control, and microvascular complications in type 1 diabetes. Diabetes Care 36(6):1629–1634 (June)
12. Mallone R, Mannering SI, Brooks-Worrell BM, Durinovic-Bello I, Cilio CM, Wong FS, Schloot NC (2010) Isolation and preservation of peripheral blood mononuclear cells for analysis of islet antigen-reactive T cell responses: position statement of the T-Cell workshop committee of the Immunology of diabetes society. J Transl Immunol 163:33–49
13. Mueller SC, März R, Schmolz M, Drewelow B (2012) Intraindividual long term stability and response corridors of cytokines in healthy volunteers detected by a standardized whole-blood culture system for bed-side application. BMC Med Res Methodol 12:112–121
14. Lee JW, Devanarayan V, Barrett YC, Weiner RS, Allinson J, Fountain S, Keller S, Weinryb I, Green M, Duan L, Rogers JA, Millham R, O'Brian PJ, Sailstad J, Khan M, Ray C, Wagner JA (2006) Fit-for-purpose method development and validation for successful biomarker measurement. Pharm Res 23(2):312–328
15. O'Hara DM, Theobald V, Egan AC, Usansky J, Krishna M, TerWee J, Maia M, Spriggs FP, Kenney J, Safavi A, Keefe J (2012) Ligand binding assays in the 21st century laboratory: recommendations for characterization and supply of critical reagents. AAPS J 14(2):316–328
16. Geist BJ, Egan AC, Yang T, Dong Y, Shankar G (2013) Characterization of critical reagents in ligand-binding assays: enabling robust bioanalytical methods and lifecycle management. Bioanalysis 5(2):227–244
17. Khan M, Bowsher R, Cameron M, Devanarayan V, Keller S, King L, Lee J, Morimoto A, Rhyne P, Stephen L, Wu Y, Wyant T, Lachno R (2015) Recommendations for adaptation and validation of commercial kits for biomarker quantification in drug development. Bioanalysis 7(2):229–242

Website References

18. www.clinicaltrials.gov/ct2/about-studies/glossary#interventional-study
19. www.fda.gov/scienceresearch/specialtopics/runningclinicaltrials
20. www.accessdata.fda.gov/scripts/cdrh/cfdocs/cfcfr/cfrsearch.cfm?cfrpart=58
21. www.cdc.gov/clia

Author Biography

Paul Rhyne obtained his Ph.D. in Cellular Immunology from the University of Tennessee at Memphis where he studied the cellular interactions between B cells and T cells. Dr. Rhyne gained post-doctoral experience in Virology at St. Jude Children's Research Hospital focusing on the tumorgenicity of Epstein-Barr Virus. Following this, Dr. Rhyne began a career in industry working in a start-up biotechnology company focused on early cancer detection technologies. He subsequently worked in the commercial antibody industry where he developed a Luminex-based product line to measure phosphorylated proteins. He joined Bristol-Myers Squibb pharmaceutical company where he built and managed a clinical biomarker group. This biomarker group was responsible for developing and validating clinical biomarker assays for BMS clinical trials. Dr. Rhyne continued to expand his career in the contract research organization industry where he was responsible for the large molecule operations at Tandem Labs. This large molecule group provided services to pharmaceutical companies that included the development and validation of clinical biomarkers, pharmacokinetic assays, and Immunogenicity assays. Recently, Dr. Rhyne joined Quintiles bioanalytical services in Marietta Georgia where he is responsible for method development of large molecule pharmacokinetic and Immunogenicity assays. In summary, Dr. Rhyne has developed hundreds of both commercial and clinical biomarker assays for the pharmaceutical industry and has successfully used many of these assays in clinical trials.

Perspectives on Tissue Biobanking for Personalized Medicine

Uma Prabhakar

Abstract The increased use of biospecimens/biobanks to address selection of specific therapies for individual patients based on their genomic and proteomic profiles, has led to the development of several issues in the conduct of biomarker and translational research. While there are significant benefits to patients, the increased utilization of biobanks for biomarker research has raised ethical, technical, regulatory, and operational considerations that require special attention by the biomarker/translational research community. Researchers pursuing such efforts should be aware of these complexities and use biospecimens appropriately to drive new directions and therapies for human diseases in various therapeutic areas.

Keywords Biospecimen · Biobank · Informed consent · Personalized medicine

Introduction

The potential of personalized medical therapy in improving treatment regimens for patients with life-threatening diseases that are not susceptible to standard of care treatments are well recognized. Therapy selection for individual patients based on the levels of specific biomarkers in the patient's blood and tissues is already on the rise. The impact of the new treatment paradigm of personalized medicine is also apparent from the surge in efforts to identify diagnostic, prognostic, mechanism of action, and early response markers relatively early in the drug development process. Typically, the resulting data and hypotheses is then tested in Phase 1 clinical trials and further validated in subsequent clinical trials for use in patient stratification and/or for monitoring drug efficacy. In the last 5–10 years, the emergence of approaches for large-scale high throughput biomarker detection and assessment through "omic's" technologies, the development of bioinformatics capabilities, specifically the systems biology efforts and, finally, the emergence of biobanking of

U. Prabhakar (✉)
BRIM Biotechnology Inc, Taipei, Taiwan, People's Republic of China
e-mail: uma.prabhakar@brimbiotech.com

R. Weiner and M. Kelley (eds.), *Translating Molecular Biomarkers into Clinical Assays*, AAPS Advances in the Pharmaceutical Sciences Series 21,
DOI 10.1007/978-3-319-40793-7_7

human tissues as a dedicated activity in several hospitals and clinics, has resulted in a major shift in the discovery-to-delivery pipeline of healthcare research.

This article provides (1) an overview of types of biobanks, (2) an overview on general perspectives relating to biobanking of human tissues for biomarker discovery in the area of drug discovery, (3) testing including informed consent operational considerations anticipated during the implementation of pharmacogenomic research, especially in clinical trials, (4) current perspectives on some of the legal and regulatory challenges when using biobank materials both in the US and internationally and, (5) issues and importance of harmonizing biobanking procedures to support international biobank interoperability and specimen sharing.

What Is a Biobank?

The most agreed upon definition of 'biobank' refers to an organized collection of human biological materials and associated information stored for one or more research purposes [1]. Human tissues may include normal and pathological solid tissues, bodily fluids (blood, blood products, urine, saliva, CSF, tear fluid, etc.) and exfoliated cells such as cell swabs, bronchial, or bladder lavage. As such, biobanks of human samples could include disease-oriented biobanks, population-based biobanks, and twin cohort studies [2, 3].

Biobank collection models can be, (a) prospective collection model, where samples are collected to meet investigators' specific requirements, (b) banking model, where samples of potential interest are collected and stored until needed and, (c) clinical trial model, where samples are collected from a clinical trial for specific or exploratory analyses [4]. Each of these models has its advantages and shortcomings. Investigators may or may not receive exactly what is requested and certain types of analyses may or may not be possible using original informed consents. In general however, biospecimens are vulnerable to environmental and biological factors introduced by routine handling during collection, processing, storage, and transport. It is therefore important that investigators are aware of the nature and types of handling variables and how these may change or influence the molecular properties of the biospecimen before it is analyzed. There is likelihood that the molecular changes may be misread as representative of the patient's disease rather than an artifact of the handling process. In fact, the lack of standard and uniform operating procedures for collection, processing, annotation, storage, and transport of biospecimens is well recognized and has resulted in a critical shortage of these important resources. Moreover, poor tissue storage conditions and lack of tissue traceability are other serious realities.

Biospecimen Research

Human tissues have been collected and stored at various institutions in the United States for over 100 years. These institutions or repositories were established to meet a specific set of objectives, and their design was integrally linked to those objectives. This resulted in tissue collection, processing, and storage techniques, etc., being extremely variable. The variability highlighted the importance of the availability of appropriately collected, consented, and annotated tissues as a critical barrier to developing genomics- and proteomics-based therapies. Until recently, majority of biobanks have been in support of oncology-based research initiatives. However, this has changed and now there are ongoing efforts in other therapeutic areas as well to develop prognostic/predictive biomarkers for similar reasons, thus broadening the scope and need to establish national, pre-competitive, regulatory compliant, and genetic-privacy protected, standardized, inclusive, highest quality network of biological sample(s) banks. In recent years, biobanks of human tissues have evolved from small-scale collections of pathological materials into structured resource centers for acquisition, storage, processing, and usage of high-quality biospecimens for research.

Several years ago, the National Cancer Institute (NCI) of the United States National Institutes of Health (NIH) undertook an intensive review process to understand the state of its funded biospecimen resources and the quality of biospecimens used in cancer research and established the Office of Biorepositories and Biospecimen Research (OBBR) in 2005 to ensure that the human biospecimens and associated data are consistent and collected according to standardized methods in order to prevent spurious analytical results that can lead to artifacts being interpreted as valid findings. Similarly, several international institutions have also taken the initiative to develop and publish best practices, which include technical recommendations for handling biospecimens as well as recommendations for ethical and regulatory practices in biobanking [5].

As shown in Fig. 1, numerous factors contribute to the quality and viability of biospecimens both in situ and also after it is acquired from the individual. Variables in handling and shipping, storage, transportation, and analytical methods can result in inconsistent and confusing data.

Awareness of what these variables are and how they may contribute to the resulting data is important in data interpretation and consistency. Comprehensive annotation of specimens containing detailed information about the donors in terms of medications, surgical procedures, pathological status, etc., should accompany each and every sample while maintaining the anonymity of the individual.

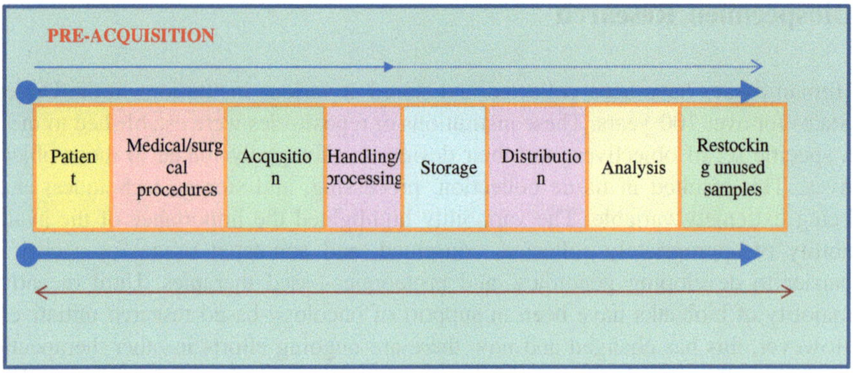

Fig. 1 Lifecycle of the biospecimen. Adapted from Office of Biorepositories and Biospecimen Research (BBRB; formerly known as OBBR): http://biospecimens.cancer.gov/http://biospecimens.cancer.gov/researchnetwork/lifecycle.asp

Biobanking Management Best Practices

The benefit that biospecimens provide to advance new therapeutic avenues at both the basic and clinical research level is evident. As discussed in a recent review [5], biobanks that house these biospecimens must have appropriate and accountable management and operations in place before they can be used for research. These include but may not be limited to, a defined infrastructure platform, trained and skilled personnel, management hierarchy with specific roles and responsibility to ensure accountability and, oversight committees. It is the responsibility of the oversight committees to ensure that the facilities have adequate space, equipment and, supplies that meets the highest standards. Best practices also include, informatics infrastructure to assist in the day-to-day operations of (1) maintenance of inventory, (2) tracking, (3) collection and data analysis and, (4) other operational procedures. Best practices will also facilitate quality assurance and quality control (QA and QC) procedures and policies, for maintaining equipment, personnel training records, and other relevant compliance associated requirements, including computer and web-based interfaces.

Biobank Ethics and Informed Consent

One of the key responsibilities for biobanks is to ensure the protection of donor information against any potential misuse for research and/or other purposes. In this context, 'Informed Consent' is a key in documenting that donors have given written permission for the use of their samples. The major concern and risk is that in biobanking and non-interventional research, personal data could end up in the wrong hands (e.g., insurance companies, employers, etc.). In recent times, there is growing awareness among patients and donors about the right to know and to control what is being done to their tissues once they are removed from their bodies and who should profit from such research. In fact, the New York Time best seller by Rebecca Skloot on the Immortal Life of Henrietta Lacks [6] has put a face on these issues. Although the debate on the ethics of biobanking has focused on obtaining general or broad consent to cover all aspects relevant to the donors choice [7], it is acknowledged that the more 'general' the informed consent is the 'less informed' it becomes. This in itself is somewhat of a challenge.

The key consideration of best practices is to consistently ensure that the rights and welfare of participants is always protected, the participants are respected and that ethically responsible research is promoted and conducted [5]. Crafting best practices in support of these purposes can be challenging. Guidelines for key issues to consider in development of governance and legacy plans as well as informed consent elements relevant to biospecimen research in addition to general requirements for informed consent under 45 CFR 46.116 have been reviewed elsewhere [5]. Additionally, the United States Human and Health Services (HHS) is also involved in the improvement of the governance plans on informed consent elements through the development of guidelines on 'custodianship' which is defined as the "caretaking responsibility for biospecimens that extends from collection through research use."

In the United States, obtaining informed consent is considered mandatory for any federally funded study or research designed to obtain scientific data either through intervention or through direct interaction with an individual. This requirement is further extended even if an investigator has to obtain any identifiable private information about the research participant [8]. However, depending upon the type of study being conducted or where in the world it is being conducted, the informed consent requirements may be subject to changes that are also influenced between different countries. Thus, there is no single, universal informed consent that can be used globally [9]. It is recommended that the scope of the informed consents remain general and yet stay 'informed' so as to anticipate and accommodate any potential future work that may be done on the biospecimens, both in terms of what kinds of determinations that will be made in the future as well as the methodology platform(s) that may be used. To accommodate the 'generality aspects', the National bioethics advisory commission recommends that the participant be offered with a 'tiered consent' with several options to choose from [10]. Biobanks or services implementing such options should ensure that the consents are structured carefully so that the interpretation and implementation of the participant choices are honored as they were made at the time of signing the consent. It is appropriate to anticipate the types of research that may be conducted on the biospecimens at a later time and describe upfront the oversight and governance procedures established at the biobank that will not only ensure best practices for ethical review and privacy protections, but will also make the informed consent less general [11].

Global Governance, Regulatory and Management Challenges

As recognized earlier in this chapter, the concept and utilization of biobanks is a major driver in the delivery of the so-called '*Personalized Medicine*' in the twenty-first century. Biomarker discovery, validation, and implementation in prevention, diagnosis, or therapy has become the cornerstone of biobanks which are no longer confined to the developed world. Given the global nature of both communicable and non-communicable diseases, the number of biobanks around the globe has increased dramatically in recent years. The harmonization of these global biobanks is often associated with significant challenges, including technical standardization, quality control, ethics, governance, regulations, and sharing of specimens and data. Country governing policies on the accessibility of these biobanks have a far-reaching influence on research, diagnosis, and treatment of diseases. The global biobanks community (both commercial and noncommercial) should identify these major challenges and address them such that gaps between different countries are minimized.

A major bottleneck in several countries, including India and China, include restrictive regulations prohibiting/limiting the export of human tissue samples because of the fear of contaminating the genetic material of the population. Although these restrictions guide each country's policies, there might be opportunities to develop flexible mechanisms and protocols that allow further health research that might benefit these countries. However, significant challenges still remain in the harmonization of structures, technical standards, and management governance despite the need for universality [11] all of which contribute to other major bottlenecks. Efforts to improving governing policies are becoming evident in countries including Italy, The Netherlands, Canada, Denmark, India, and developing countries [12–17] and it is anticipated that this trend will continue in the coming years.

Overcoming Challenges Through Collaborations and Partnerships

Yassin et al. [18] in a recent publication have provided a good perspective and framework for the biospecimen custodianship. Given varying interpretations and complexities, custodianship is often misunderstood and it is deemed essential that there exist structured and transparent policies, which ensure the right to control the use and distribution of biospecimens. Similarly, the collection of high-quality biospecimens appropriately consented and thoroughly annotated, requires the cooperation of surgeons, pathologists, nursing staff, researchers, and patients/donors. Standardization across biobanks through national and international collaborations and partnerships will improve biospecimen quality and patient care.

NCI promotes such partnerships that include, working with the *College of American Pathologists* (*CAP*) to develop evidence-based SOPs for biospecimens. Similarly, working with *Cancer Biomarkers Collaborative* (*CBC*) that includes representatives from the Food and Drug Administration (FDA), the American Association for Cancer Research, the pharmaceutical industry, academia and patient groups, ensures the advancement of the Critical Path Initiative (CPI). Further, to achieve harmonization of best practices in this area, a number of organizations are attempting to resolve key issues internationally. The International Society for Biological and Environmental Repositories (ISBER) [19] hosts annual meetings and through its working groups, best practices and other resources provides a forum for sharing biobanking information. Similarly, European, Middle Eastern and African Society for Biobanking and Biopreservation (ESBB) were created to provide similar coordination and educational support on a regional basis [20]. It is anticipated that such efforts will help develop SOPs for addressing practical, ethical, and legal challenges and ensue high-quality biospecimens.

Chapter Summary

1. General perspectives relating to biobanking of human tissues for biomarker discovery in the area of drug discovery and testing.
2. Informed consent operational considerations anticipated during the implementation of pharmacogenomic research, especially in clinical trials.
3. Types of biobanks and, current perspectives on some of the challenges, both legal and regulatory when using biobank materials both in the US and internationally.
4. Issues and importance of harmonizing biobanking procedures to support biobank interoperability and specimen sharing.
5. Bottlenecks of international biobanking.

Acknowledgment I gratefully acknowledge the support and advice provided by the late Dr. Mahadev Murthy in the preparation of this manuscript and dedicate this book chapter in his memory. Dr. Murthy's major interests included biomarkers and their potential utility in translational and personalized medicine including, prediction, diagnosis, and treatment of diseases.

References

1. Kauffmann F, Cambon-Thomsen A (2008) Tracing biological collections: between books and clinical trials. JAMA 299:2316–2318
2. Gottweis H, Zatloukal K (2007) Biobank governance: trends and perspectives. Pathobiology 74:206–211
3. Yuille M, van Ommen G-J, Brechot C et al (2008) Biobanking for Europe. Brief Bioinform 9:14–24

4. Hewitt RE (2011) Biobanking: the foundation of personalized medicine. Curr Opin Oncol 23:112–119
5. Vaught J, Lockhart NC (2012) The evolution of biobanking best practices. Clin Chim Acta 413:1569–1575
6. Skloot R (2010) The immortal life of Henrietta Lacks, 1st edn. Crown, New York
7. Cambon-Thompson A, Rial-Sebbag E, Knoppers BM (2007) Trends in ethical and legal frameworks for the use of human biobanks. Eur Respir J 30:373–382
8. Belmont report, ethical principles and guidelines for the protection of human subjects of research. The National Commission for the Protection of Human Subjects of Biomedical and Behavioral Research; April 18. http://www.hhs.gov/ohrp/humansubjects/guidance/belmont.html (1979)
9. Ricci DS, Franc M (2008) Implementation of pharmacogenomic sample collection in clinical trials. In: Cohen N (ed) Pharmacogenomics and personalized medicine. Humana Press, p 27
10. National Bioethics Advisory Commission (1999) Research involving human biological materials: ethical issues and policy guidance, p 67
11. Smith ME, Aufox S (2013) Biobanking: The melding of research with clinical care. Curr Genet Med Rep 1:122–128
12. Bravo E, Napolitano M, Santoro F, Belardelli F, Federici A (2013) The Italian hub of population biobanks as a potential tool for improving public health stewardship. Biopreserv Biobank 11(3):173–175
13. Calzolari A, Napolitano M, Bravo E (2013) Review of the Italian current legislation on research biobanking activities on the eve of the participation of national biobanks' network in the legal consortium BBMRI-ERIC. Biopreserv Biobank 11(2):124–128
14. Brimo F, Aprikian A, Latour M, Têtu B, Doueik A, Scarlata A, Hamel L, McKercher G, Saad F, Lacombe L, Carmel M, Chevalier S (2013) Strategies for biochemical and pathologic quality assurance in a large multi-institutional biorepository; the experience of the PROCURE Quebec Prostate Cancer Biobank. Biopreserv Biobank 11(5):285–290
15. Hewitt R, Hainaut P (2011) Biobanking in a fast moving world: an international perspective. J Natl Cancer Inst Monogr 42:50–51
16. Artene S-A, Ciurea ME, Purcaru SO, Tache DE, Tataranu LG, Lupu M, Dricu A (2013) Biobanking in a constantly developing medical world. Sci World J 2013(343275):1–5.
17. Hall MA, King NMP, Perdue LH, Hilner JE, Akolkar B, Greenbaumd CJ, McKeonc C, the T1DGC (2010). Biobanking, consent, and commercialization in international genetics research: the Type 1 Diabetes Genetics Consortium. Clin Trials 7:S33–S45
18. Yassin R, Lockhart N, González del Riego M et al (2010) Custodianship as an ethical framework for biospecimen-based research. Cancer Epidemiol Biomarkers Prev 19(4): 1012–1015.
19. http://www.isber.org.
20. http://www.esbb.org.

The Role of Commercial Biomarker Assay Kits in Preclinical and Clinical Trials

William Nowatzke and Ronald R. Bowsher

Abstract The aim of this chapter is to discuss the role of commercial assay kits when quantitatively measuring biomarkers to support preclinical and clinical drug development programs. Research use only and clinical diagnostic kits will be compared and their usefulness addressed in the context of the 2013 FDA draft guidance on bioanalytical method validation. Analytical method performance characteristics will be discussed as well as mechanisms to generate consistent biomarker data throughout the life of the study.

Key Terms

Biomarker	A characteristic that is objectively measured and evaluated as an indicator of normal biologic processes, pathogenic processes, or pharmacologic response to a therapeutic intervention
Definitive Quantitative Assay	Uses calibrators fit to a regression model to calculate the absolute quantitative values for unknown samples and the reference standards are well defined and fully representative of the endogenous biomarker
Bioanalytical Method Validation	A formal process of evaluating the analytical performance characteristics of an assay to ensure it provides reliable data
Research Use Only kits	Commercial assay reagents and components intended to be used to generate research quality data

W. Nowatzke (✉)
Radix BioSolutions, 111 Cooperative Way, Ste 120, Georgetown, TX 78660, USA
e-mail: bill.nowatzke@radixbiosoltions.com

R.R. Bowsher
B2S Consulting, 6656 Flowstone Way, Indianapolis, IN 46237, USA
e-mail: ronb@b2s-stats.com

R. Weiner and M. Kelley (eds.), *Translating Molecular Biomarkers into Clinical Assays*, AAPS Advances in the Pharmaceutical Sciences Series 21,
DOI 10.1007/978-3-319-40793-7_8

Clinical Diagnostic Kits	FDA approved reagents and components intended to be used for the diagnosis of human disease and medical risk factors

Introduction

It is well recognized by the pharmaceutical industry that molecular biomarkers (often peptides or proteins) measured in biological matrices may be used to accelerate candidate drug advancement by indirectly detecting target engagement (i.e., proximal biomarker), [1] and desired physiological changes to therapeutic interventions [2, 3]. Alternatively, the unnecessary waste of company resources and redirection to alternative therapeutic drug candidates may be benefited by the 'fail-early' concept. The FDA has addressed the increasing importance of biomarkers in drug development programs indicating that '…the two most important areas for improving medical product development are biomarker development (Topic 1) and streamlining clinical trials (Topic 2)' [4, 5].

It is important to emphasize the notion that analytical data generated to support a drug development program is only as reliable as the analytical validity of the bioanalytical method used to quantitate the biomarker concentrations. While this concept is understood well by laboratorians (i.e., 'biomarker data generators') who are tasked with developing and validating the analytical methods, it may not be fully appreciated by other customers (i.e., 'biomarker data consumers') who are eager to access the biomarker data to help advance their drug development programs.

Several excellent publications precede this chapter, including recommendations on developing and validating bioanalytical methods to measure biomarkers [6–12] and the use of research use only and diagnostic kits to support biomarker measurements in the context of drug development programs [13–15].

The aim of this chapter is to discuss biomarker commercial kit options in conjunction with the minimum analytical performance expectations that should be explored when supporting preclinical and clinical drug development programs. This topic is timely, as regulatory bodies are considering applying rigorous method validation and performance characterization to biomarker methods that were previously defined solely for data generated to support specific pharmacokinetic studies [16, 17]. Speculation on why regulatory agencies are more closely scrutinizing biomarker data now, compared to fifteen years ago, could include the evolution of personalized medicine/companion diagnostics and the use of surrogate biomarkers to define clinical endpoints for trials investigating chronic diseases. In the case of the former, physicians are making healthcare decisions based upon biomarker concentrations (or the presence/absence of a macromolecule). Regarding the latter, the use of a surrogate biomarker may permit the conductance of clinical studies that would otherwise have been fiscally or ethically impractical based on a clinical outcome endpoint. The importance of this data emphasizes the need for analytical methods that generate quality data.

The scope of this chapter is limited to discussions on quantitative assays [7]. Definitive quantitative assays use calibrators fit to a regression model to calculate the absolute quantitative values for unknown samples. Furthermore, reference standards are well defined and fully representative of the endogenous biomarker [9]. Relative quantitative assays are similar to definitive quantitative assays with the exception that the reference standards may not exactly represent the endogenous analyte of interest.

Commercial Assay Kits

Commercial assay kits do have some drawbacks as discussed below, however, these kits are popular because they are convenient to use and liberate laboratory resources from conducting method development. Kits are available for the immediate measurement of many thousands of biomarker molecules. These assays are designed to be used on popular, industry-standard platforms, and may be ordered in bulk or large quantities to maintain consistency for lengthy studies. Generally, there are two kit options; (1) research use only (RUO) kits, and (2) FDA approved clinical diagnostic kits.

Research Use Only Kits

As implied by the term, RUO kits are developed for research purposes with a commercial vendor strategy to offer an exhaustive menu of analytes and to be first to market. These kits are marketed without any implied guarantee as to the quality of performance. Kits are available from a multitude of sources, and there is no standardization within the industry in terms of the type or extent of method development and the validatability of the assay. It is important to also recognize that RUO kits are designed by the manufacturer to be suitable for measuring biomarkers across a broad range of applications often without consideration of the type of sample matrix. Changes in manufacturer lot numbers often result in a change in the performance of the assay. It is not uncommon to observe shifts in instrument responses, corresponding to changes in interpolated sample and quality control results. These kits are not intended to be used to generate data that will be reviewed by regulatory agencies without modifications to the method that bring them to an industry-standard validation state. That being said, some RUO kits perform exceptionally well, and may be relied on to immediately screen samples for biomarkers of interest.

When evaluating biomarkers in support of drug development, precision tends to be a more critical assay characteristic than accuracy, as one is usually trying to detect changes in biomarker concentrations in response to a therapeutic intervention, rather than an absolute value. Optimization of sample analysis work flow can minimize imprecision not directly attributed to the analytical method. For example, avoid dispersing samples from an individual in different analytical runs or on different assay plates, and attempt to use the same kit lot number throughout the evaluation. If the evaluation is found to be acceptable, the assay should be modified and validated to an industry standard [9] prior to generating data that may be submitted to regulatory agencies. One set of validation recommendations cannot define the needs of every possible biomarker study, therefore, the minimum validation characteristics should be evaluated, but a fit-for-purpose approach is recommended (Table 1).

Table 1 Comparison of bioanalytical method performance characteristics for biomarker kits

Characteristic	RUO kits[a]	Fit-for-purpose	Diagnostic kit[b]	FDA draft guidance
Dynamic range (LLOQ, ULOQ)	No	Yes	Yes	Yes
Sensitivity	No	Yes	Yes	Yes
Curve fitting	No	Yes	Yes	Yes
Selectivity	No	Yes		Yes
Specificity	No	Yes	Yes	Yes
Parallelism	No	Yes	No	Yes
Dilutional linearity	No	Yes	Yes	Yes
Precision and accuracy (analytical)	No	Yes	Yes	Yes
Relative accuracy/recovery (biological)	No	Yes	No	Yes
Robustness	No	Yes	No	Yes
Sample handling and collection, processing, storage and analyte stability	No	Yes	Yes	Yes
Reportable range	No	Yes	Yes	Yes
Reference interval	No	No	Yes	No
Biomarker clinical validity demonstrated	No	No	Yes	No
Quality control	Variable, buffer based	Mimic samples	Commercial	Mimic samples

[a]Manufacturer validations are not standardized. The degree of verification is guided by the package insert and the intended use of the study data
[b]Some performance characteristics terms may vary for CLIA validations. *CLIA* Clinical Laboratory Improvement Amendments; *LLOQ* Lower limit of quantitation; *ULOQ* Upper limit of quantitation

Clinical Diagnostic Kits

Until recently, the intended use of clinical diagnostic kits has been to guide patient care. This includes identification of health risk factors, diagnosis of disease or acute events, and physiological responses to medical interventions. Unlike commercial RUO kits, clinical diagnostic kits must undergo FDA approval. In addition to demonstrating that the measured analyte has clinical validity, manufacturing is controlled following the medical devices quality system regulation (21 CFR Part 820). These kits are manufactured under good manufacturing practices (GMP) and the critical reagents are regulated—"Analyte specific reagents (ASR's) are antibodies, both polyclonal and monoclonal, specific receptor proteins, ligands, nucleic acid sequences, and similar reagents which, through specific binding or chemical reaction with substances in a specimen, are intended for use in a diagnostic application for identification and quantification of an individual chemical substance or ligand in biological specimens" (21 CFR Part 864). When utilizing these kits to analyze patient samples for diagnostic purposes, CLIA regulations apply [18]. A detailed summary of CLIA regulations is not possible in this chapter, however, major differences from laboratories supporting drug development are that CLIA laboratories must be accredited (licensed), must perform blind sample proficiency challenges, are subject to scheduled regulatory inspections, and must be operated by a medical doctor or board certified Ph.D. director.

In addition to the stringent manufacturing conditions, clinical diagnostic kits undergo an extensive analytical method validation following standards defined by the Clinical and Laboratory Standards Institute (CLSI). CLSI is a nonprofit laboratory medicine organization that facilitates a unique process of developing clinical laboratory testing standards based on input from and consensus among industry, government, and healthcare professionals. However, these method validations differ significantly from recommendations to validation biomarker assays to support drug development [16, 17]. The three most critical differences are (1) the use of a buffer-based assay system, rather than biological matrices—a consequence of using very small sample volumes (often <5 μL) on automated clinical analyzers, (2) defining the analytical range and sensitivity as the lowest value that is significantly different than zero, rather than a lower limit of quantitation (LLOQ), and, because instruments are FDA approved, calibrator samples may not be evaluated with every analytical run.

The above manufacturing measures are far more comprehensive than RUO commercial kits or most biomarker assays that are developed and used in-house. As a consequence, laboratories supporting drug development programs are expressing an interest in utilizing clinical diagnostic kits to measure biomarkers. However, there are several obstacles that must be overcome before these kits may be used (Table 2).

When considering the purchase of a kit to support a biomarker study, a systematic approach should be taken to identify the kit that is most suitable for the intended purpose. Identifying the appropriate kit for a given study is critical, as

Table 2 Advantages and disadvantages of using Clinical Diagnostics Kits to support drug development programs

Advantages
FDA Approved
Extensive method development and CLSI validation
Continuous assay performance monitoring by extensive reference laboratory networks
Demonstration of clinical biomarker validation
GMP Manufacturing and traceable documentation
Lot-Lot Consistency
Potential for high-throughput analyses
Cost per sample may be reduced
Kits costs often comparable to RUO kits
Disadvantages
Calibrators and quality controls are often buffer based
Number of calibrators may not meet regulatory recommendations
Sample volume optimized for automated clinical analyzers (<5 μL)
Packaging may be intended for placement on clinical analyzer—limited access to reference standard
CLSI method validation does not meet regulatory agency standards for supporting drug development programs—intended for diagnostic purposes
Analytical range/sensitivity defined by limit of detection rather than lower limit of quantitation
May require specialized analyzer with closed software systems
Application have a narrow scope and limited sample types
Novel biomarker methods not available
The method has likely not been evaluated for interference from the therapeutic or disease state matrix components

significant financial and staff resources may be invested in a given project. A commercial kit should be selected while considering the intended use of the biomarker data. A RUO kit may be appropriate for screening gross differences in archived samples and demonstrating feasibility, however, a validatable clinical diagnostic kit may be more suitable for confirming a proof of concept study. An internal review of the historical significance and measurement of the biomarker is recommended, as well as consulting with external colleagues and key opinion leaders. Literature and patent search may reveal commercial kits that were successfully used in similar projects. It is also worthwhile to investigate the source of commercial assays that are being utilized by contract research organizations and central reference laboratories. Upon identifying possible kits, package inserts can usually be obtained electronically. Inserts will provide the assay platform information, the assay design, a summary of the assay performance by the vendor, the approximate range and sensitivity of the method, and often literature references or poster abstracts in which the assay was successfully used. Contacting the kit manufacturer, technical support, or sales representative may yield additional information that has not been publicly released.

Evaluating Commercial Kit Performance

Currently published recommendations outline the best practices for validating biomarker assays dates just over a decade [6]. Many of these early discussions were focused on bioanalytical methods that were developed in-house to measure novel biomarkers. A follow up white paper has been embraced by industry as the standard for the fit-for-purpose validation of definitive quantitative biomarkers [9].

The bioanalytical community is delighted that the FDA is beginning to address expectations for the analytical validation of bioanalytical methods '…to measure in vivo biomarker concentrations in biological matrices, such as blood or urine… when biomarker data will be used to support a regulatory action, such as the pivotal determination of safety and/or effectiveness or to support labeled dosing instructions…' [19]. It is stated in this document that the minimum method characteristics that should be evaluated for a biomarker method are accuracy, precision, selectivity, range, reproducibility, and stability (Table 1). Method validation should address the same questions as pharmacokinetic assays; however, it is ultimately up to the sponsor to incorporate the extent of method validation that they feel is appropriate. Although the 2013 Guidance is a draft, a critical meeting was held in December 2013 during which time the draft Guidance was extensively reviewed, discussed, and critiqued by colleagues in industry, CROs, academia, FDA, and international regulatory representatives from other countries [20]. It was agreed that incorporation of content to discuss biomarker method validation and the use of diagnostic kits—into what has traditionally been a pharmacokinetic bioanalytical guidance, was a first step towards ultimately defining the best practices for validating these methods to met regulatory expectations. A subsequent workshop (Crystal City VI) was held in September 2015. It is anticipated that a conference report will be published in the very near future.

It is highly recommended that sponsors continue to consult with the appropriate regulatory agencies to discuss expectations specific to their program.

Maintaining a Consistent Assay Over Time

Regardless of the study type or analyte, one of the most critical issues when providing bioanalytical support to drug development programs is the ability to prospectively generate reliable results as a product of having consistent assay performance. The source of changes in assay performance can be the result of several factors. Robust ligand binding assays and LC-MS/MS methods tend to be less prone to small differences in analytical technique. Likewise, liquid handling automation can reduce assay-to-assay differences. However, there are cases where changes in assay performance are observed when performed by two different analysts. A second source of assay shift is related to the management of critical reagents. Capture antibodies/receptors and conjugated detection reagents are typically prepared in manufacturing lots, that are prone to the generation of reagents at different degrees of purity, concentration, conjugation efficiency, and extent of protein denaturation. Additionally, manufacturers occasionally change the source of the some critical reagents (e.g., reference standards) or the packaging process (e.g., lyophilized solid to frozen in solution). Laboratory conditions have been known to cause differences in assay performance, and it is recommended that temperature and humidity be monitored. These charts are useful for troubleshooting, rather than defining assay run conditions. Assay drift may occur as a consequence of deteriorating reagents, instrument component malfunctioning or analyte instability.

It is recommended that matrix-based bridging quality controls be used to monitor and detect shifts and trends in commercial assay kits that may be a consequence of the above issues. If a change is detected, the use of incurred samples to evaluate assay reproducibility should be part of the investigation. The laboratory should monitor not only QC concentrations, but also instrument responses for blanks, standards, and quality controls. Comparison of this data to historical assay performance is invaluable for troubleshooting and identifying the root cause of assay inconsistency.

Once a change in assay performance is detected and investigated, the laboratory must work with the vendor to confirm the observation. The laboratory has limited control of the resolution at this point. For RUO kits, the probable outcome is that the change is acknowledged, but the vendor has no solution for individual laboratories. FDA approved clinical diagnostic kits may have recalls, or may have already addressed the shift by changes in calibrator assignments and instrument programming measures.

Solutions at the laboratory level may include applying a correction factor to data generated with the new kit or reanalysis of previously analyzed samples. When feasible, the best approach is to anticipate reagent needs for the life of the study. Efforts should be made to work with the vendor to provide bulk reagents or kits with appropriate expiration dating. An excellent example of such an approach is described by Lee [21].

Conclusions

- Commercial assay kits can offer a convenient alternative to internally developed lab tests.
- Samples may be analyzed immediately on industry-standard platforms and in the case of clinical diagnostic kits, will have undergone extensive development, analytical, and clinical validation.

- However, the laboratory sacrifices control over reagent management and the design of the performance characteristics of the method. Furthermore, assays used to generate data that will be reviewed by regulatory agencies are likely to not meet the anticipated expectations for validated biomarker assays that are used to support drug development programs.
- Thus, the commercial assay kit is the starting point for method optimization, leading to matrix-based method validations.
- The method validation process is expected to follow a validation process similar to a traditional pharmacokinetic method validation. However, because of unique challenges specific to biomarker bioanalysis (e.g., acquiring a reference standard that is identical and equally immunoreactive to the biological counterpart), a creative fit-for-purpose approach may complement recommendations in regulatory guidance documents.
- Finally, shifts in the behavior of commercial assay kits should be anticipated, especially when manufacturing components change.

References

1. Krishna R, Herman G, Wagner JA (2008) Accelerating drug development using biomarkers: a case study with Sitagliptin, a novel DPP4 inhibitor for type 2 diabetes. AAPS J 10(2):401–409
2. Colburn WA (2003) Biomarkers in drug discovery and development: from target identification through drug marketing. J Clin Pharmacol 43(4):329–341
3. Williams SA, Slavin DE, Wagner JA, Webster CJ (2006) A cost-effectiveness approach to the qualification and acceptance of biomarkers. Nat Rev Drug Discov 5(11):897–902
4. Woodcock J (2009) Chutes and ladders on the critical path: comparative effectiveness, product value, and the use of biomarkers in drug development. Clin Pharmacol Ther 86(1):12–14
5. US FDA. Innovation or stagnation? Challenge and opportunity on the critical path to new medical products. www.fda.gov/ohrms/dockets/ac/04/briefing/4039b1_03_Critical%20Path%20whitepaper.pdf
6. Lee JW, Smith WC, Nordblom GD, Bowsher RR (2003) Validation of assays for the bioanalysis of novel biomarkers. In: Bloom JC, Dean RA (eds) Biomarkers in clinical drug development. Marcel Dekker, NY, USA, pp 119–149
7. Bowsher RR (2003) Analytical validation of assays for novel biomarkers. Presented at AAPS Workshop, Method validation and measurement of biomarkers in nonclinical and clinical samples in drug development. Salt Lake City, UT, USA
8. Lee JW, Weiner RS, Sailstad JM et al (2005) Method validation and measurement of biomarkers in nonclinical and clinical samples in drug development: a conference report. Pharm Res 22:499–511
9. Lee JW, Devanarayan V, Barrett YC et al (2006) Fit-for-purpose method development and validation for successful biomarker measurement. Pharm Res 23:312–328
10. Chau CH, Rixie O, McLeod H et al (2008) Validation of analytical methods for biomarkers used in drug development. Clin Cancer Res 14(19):5967–5976
11. AAPS Workshop. Biomarker assay development and validation: challenges and creative solutions associated with emerging biomarker assays. Georgia World Congress Center, Atlanta, GA, USA, 15–16 Nov 2008
12. Lee JW, Hall M (2009) Method validation of protein biomarkers in support of drug development or clinical diagnosis/prognosis. J Chromatrogr B 877:1259–1271

13. Bowsher RR, Sailstad JM (2008) Insights in the application of research-grade diagnostic kits for biomarker assessments in support of clinical drug metabolism: bioanalysis of circulating concentrations of soluble receptor activator of nuclear factor Kb ligand. J Pharm Biomed Anal 48:1282–1289
14. Nowatzke WL, Cole TG, Bowsher RR (2010) Systematic analytical validation of commercial kits for the determination of novel biomarkers for clinical drug development. Bioanalysis 2(2):237–247
15. Khan MU, Bowsher RR, Cameron M et al (2015) Recommendations for adaption and validation of commercial kits for biomarker quantification in drug development. Bioanalysis 7(2):229–242
16. Guidance for Industry On Bioanalytical Method Validation (2001) Fed Reg 66:28526–28527
17. Viswanathan CT, Bansal S, Booth B et al (2007) Quantitative bioanalytical methods validation and implementation: best practices for chromatographic and ligand binding assays. Pharm Res 24(10):1962–1973
18. Clinical Laboratory Improvement Amendments of 1988, 42 USC 201 note, Public Law 100-578, 102 STAT.2903
19. Draft Guidance for Industry On Bioanalytical Method Validation. Federal Register 78 FR 56718, 56718–56719 (2013). AAPS Workshop. Quantitative bioanalytical methods validation and implementation: The 2013 Revised FDA Guidance. Baltimore Hilton, Baltimore, MD, USA, 03–05 Dec (2013)
20. Booth B, Arnold ME, DeSilva B et al (2014) Workshop report: Crystal City V-quantitative bioanalytical method validation and implementation: the 2013 revised FDA guidance. AAPS J 17(2):277–288
21. Wang J, Lee J, Burns D et al (2009) "Fit-for-purpose" method validation and application of a biomarker (c-terminal telopeptides of type 1 collagen) in denosumab clinical studies. AAPS J 11(2):385–394

Quantification of Protein Biomarkers Using Liquid Chromatography Tandem Mass Spectrometry

Hendrik Neubert

Abstract LC-MS/MS, particularly when linked with immunoaffinity enrichment, has emerged as a highly capable bioanalytical technique for the quantitative measurement of protein biomarkers and therapeutic proteins, thus impacting translational pharmacology. A key advantage of a protein LC-MS/MS assay over other bioanalytical techniques is the high measurement specificity that can be achieved. Immunoaffinity enrichment techniques using anti-protein or anti-peptide antibodies, or both in a sequential manner, extend LC-MS/MS assay sensitivity for protein biomarkers into the pg/mL range. Assay translation between species can be facilitated by selecting proteotypic peptides that are conserved in the same protein across species, if available, to allow the same MS detection method, the same SIL standard peptide and the same anti-peptide antibody can be used. Practical challenges to routine implementation in clinical assays are being overcome by the use of standardized workflows, liquid handling robotics, and robust LC-MS/MS configurations.

Keywords Liquid chromatography tandem mass spectrometry · Translational pharmacology · Proteotypic peptide · Selected reaction monitoring · Protein and peptide immunoaffinity

Key Terms

Translational pharmacology Investigations of drug effects on pathways and disease to establish a mechanistic link between in vitro or ex vivo to in vivo systems as well as within and between species. For example, to identify in vivo pharmacology and biomarkers in preclinical species that can also be measured in humans

H. Neubert (✉)
Biomedicine Design, Pfizer Worldwide Research and Development,
1 Burtt Road, Andover, MA 01810, USA
e-mail: hendrik.neubert@pfizer.com

R. Weiner and M. Kelley (eds.), *Translating Molecular Biomarkers into Clinical Assays*, AAPS Advances in the Pharmaceutical Sciences Series 21,
DOI 10.1007/978-3-319-40793-7_9

Proteotypic peptide	A peptide, enzymatically released from a protein as part of an LC-MS/MS assay, which serves to unambiguously identify that protein
Selected reaction monitoring	Selected reaction monitoring (SRM) is a tandem mass spectrometry technique in which an ion of specified mass-to-charge ratio is selected in the first mass spectrometry stage. One or several product ions of a specified mass-to-charge ratio resulting from fragmentation of the precursor are detected in the second mass spectrometry stage
Sequential protein and peptide immunoaffinity	Dual immunoaffinity sample preparation technique used for measuring low abundance protein biomarkers in biological matrices by mass spectrometry. Protein immunoaffinity enrichment of the protein biomarker using an anti-protein capture reagent is followed by digestion and immunoaffinity enrichment of one or several enzymatically released peptides using anti-peptide antibodies

Protein Biomarkers and the Need for Their Selective Bioanalysis

A major aim of translational pharmaceutical and biomedical research is to transition drug candidates and their targets and pathways from preclinical discovery to clinical development. Biomarkers can facilitate decisions in ***translational pharmacology***, safety as well as precision medicine for patient stratification [1]. While many classes of endogenous molecules can constitute biomarkers, the need to quantitatively measure endogenous proteins as biomarkers is undeniable. In addition, the bioanalysis of the therapeutic targets themselves, which are proteins in many cases, can be a key biomarker measurement. Target analysis in normal and disease, or following treatment, both in systemic circulation and in tissues can support ***translational pharmacology*** by assisting with the construction of the pharmacokinetic/pharmacodynamic relationship. An important extension is the analysis of target engagement by the therapeutic, which is frequently important for rationalizing the selection of the dosing regimen. Downstream pharmacodynamic biomarkers are mechanistically linked to the therapeutic target and are a direct measure of pathway modulation by the therapeutic.

Localization, abundance, internalization, or turnover rate, to name a few, are diverse attributes of the proteins investigated as biomarkers. Most commonly, the concentration of a soluble protein biomarker in a biological fluid is determined. However, depending on the question, biomarker assays are also needed to determine the amount of membrane-associated protein in solid tissues, such as in clinical biopsies. Moreover, the occurrence and abundance of posttranslational modifications such as phosphorylation, glycosylation or ubiquitination, and splice variants or the rate of protein synthesis or degradation can also be a biomarker. A constant evolution of technologies and assay strategies is required to be able to address these, often times, challenging bioanalytical questions for protein biomarkers. There are many competent technologies available for biomarker measurements, many of these are described in this book. This section will focus on quantitative protein mass spectrometry, which is still a comparatively young application area, but has matured substantially in recent years and is increasingly becoming a major player in the quantification of protein biomarkers in translational and clinical research. One key advantage quantitative protein mass spectrometry has over other bioanalytical tools is that one can achieve high measurement specificity.

From Mass Spectrometry-Based Proteomics to Quantitative LC-MS/MS Protein Biomarker Assays

Since 1990s, qualitative proteomics has focused on the identification of proteins from biological samples resulting in the creation of valuable catalogs of detectable proteins [2]. Detection is typically facilitated via digestion of proteins into peptides using specific proteases such as trypsin. Over the years, mass spectrometry has become the dominant proteomics detection tool, employing mostly quadrupole time-of-flight (QTOF) and orbitrap mass analyzers. With evolving instrumentation and advancing workflows for sample preparation, it has been possible to increase the number of proteins identified per sample from hundreds to thousands in each individual experiment including the detection of posttranslational modifications. In the last decade, many global unbiased proteomics studies have been enabled by semiquantitative workflows, allowing comparison of relative protein abundance in different samples and conditions. This facilitated a proteomics study design that could link identification of putative protein biomarkers and their relative abundance to a functional biological endpoint. To this end, data-dependent acquisition (DDA) methods are employed that are either based on label-free comparison of mass spectrometry signals or based on chemical or metabolic labeling with ^{13}C or ^{15}N, which can be distinguished by the mass spectrometer from isotopes that occur naturally in high abundance [2]. Major examples include labeling techniques such as iTRAQ, TMT, or SILAC [3]. Recent advances in data-independent acquisition (DIA) methods demonstrate not only deep proteome coverage [4], but also the ability to semiquantitate all measured proteins, and require no a priori knowledge of anticipated protein biomarker changes. Proteomics researchers also perform hypothesis-driven studies, where the experiment focuses on the detection of a known set of putative protein biomarkers. The mass spectrometer is programmed to only analyze proteins of interest, providing improved sensitivity compared to a global proteomics survey experiment where as many proteins as possible are detected. Hypothesis-driven proteomics experiments are typically done in a semi-quantitative manner by comparing relative signal intensities. Both, unbiased and hypothesis-driven experimentation can discover new protein biomarkers or signatures. These can be followed up with fully quantitative assays capable of analyzing larger sample sets with higher analytical rigor to achieve biological validation of the putative biomarker.

The proteomics field has developed numerous sample preparation techniques including digestion and enrichment approaches as well as mass spectrometry detection methods that are fundamental to today's quantitative protein LC-MS/MS assays. Methodologies used in contemporary LC-MS/MS quantification of protein biomarkers also draw on the vast experience that exists with LC-MS/MS quantification of small molecule biomarkers. This includes basic quantification concepts such as the use of stable isotope labeled standards, determination of quantitative LC-MS/MS assay performance, and how to handle endogenously detectable analyte during assay development. This has led to an ongoing discussion of how bioanalytical performance and acceptance criteria can be adapted for the quantification of protein biomarkers.

Preparing the Sample for Mass Spectrometry

In most cases, the sample containing the protein to be quantified is digested into measurable, ***proteotypic peptides*** prior to LC-MS/MS using a specific protease. Most commonly, trypsin is employed which cleaves C-terminal to arginine and lysine amino acids, except if adjacent to a proline. Other enzymes can be used as well, such chymotrypsin or endoproteases Asp-N, Lys-C, Arg-C, or Glu-C. Chemical digestion with acid [5] or cyanogen bromide [6, 7] has also been demonstrated for protein quantification. The advantage of digestion is that higher mass spectrometric sensitivity can be achieved when measuring peptides compared to proteins as instrumental sensitivity declines with increasing mass. Amino acid sequence homology of the selected peptide with relevant database entries needs to be confirmed in silico. Furthermore, when measuring peptides as quantitative surrogates of the proteins they originate from, it is important to experimentally confirm that peptide abundance is truly representative of protein abundance. This is central to quantitative protein LC-MS/MS, particularly for the analysis of protein biomarkers, where premature forms, posttranslational modifications or splice variants, or other forms resulting from biological processing can correlate with biological effect. If possible and depending on the application, additional peptides should be measured simultaneously, to ensure that results are consistent. In fact, it is a key advantage of LC-MS/MS that several peptides from the same protein can be measured simultaneously. This allows obtaining more complete amino acid sequence coverage of the protein biomarker to span multiple domains of relevance to biological function. Multiplexed quantification of several peptides originating from one or more proteins can then be incorporated into one protein biomarker assay.

Sample types for protein biomarker quantification can range from cell cultures, plasma or serum and other fluids to solid tissues. The dynamic range of protein abundance in these samples can be large, spanning >10 orders of magnitude in human serum [8]. A sample preparation workflow consisting only of protease digestion prior to conventional LC-MS/MS is typically not very sensitive which limits the application of the assay to only highly abundant proteins. This is due to a number of factors including the limited loading capacity on liquid chromatography columns and the dynamic detection range of the mass spectrometer. Therefore, in order to measure biomarker proteins of lower abundance, sample fractionation or enrichment needs to be incorporated in an assay workflow.

Fractionation

There are well-established fractionation techniques, both at protein or peptide level, which can be used in quantitative protein biomarker LC-MS/MS assays. These include fractionation based on differential solubility or hydrophobicity, molecular weight, charge, and pI. Many proteomics fractionation techniques have been used in the quantitative analysis of proteins biomarkers; however, prominent examples of methods that can be easily implemented in protein biomarker assays are solid-phase extraction (SPE), protein precipitation or ion exchange chromatography. SPE can be used as a positive or negative selection tool depending on the stationary phase and analyte characteristics [9]. Fractionation based on ion exchange can also be incorporated into a protein quantification workflow, for example at digest level using a weak cation exchange monolithic trap in an online configuration prior to LC-MS/MS [7].

Driving Sensitivity with Immunoaffinity Enrichment

Using antibodies to enrich the protein biomarker, the enzymatically released peptide or both in a sequential manner can provide tremendous gains in LC-MS/MS assay performance [10], particularly with respect to sensitivity, dynamic range and throughput. Immunoaffinity (IA) at protein level using an antibody (Fig. 1a), which is considered equivalent to a protein capture step in a ligand binding assay, allows LC-MS/MS analysis of protein biomarkers that are in the low to mid pg/mL concentration range or above in plasma or serum. Sensitivity is typically scalable with sample volume if sufficient capture antibody is used. A preferred practical implementation, which has been tested both in preclinical and routine clinical protein biomarker assays, is the use of biotinylated antibodies paired with streptavidin-coated paramagnetic beads. This workflow is well suited to operation on liquid handling robotics [7, 11–15] providing a technical solution that can be easily standardized and validated for clinical implementation. One key feature of the protein IA technique is that further analyte selection can be performed, similar to a ligand binding assay. For example, selecting a capture antibody to a specific epitope on the protein biomarker or therapeutic target, allows either intentionally competing or not competing with an endogenous binding partner, or a binding biotherapeutic, such as a monoclonal antibody. These types of measurements can be critical to developing an understanding for example of mechanistic pharmacology in pre-clinical studies and early clinical drug trials.

Another frequently employed IA strategy is the use an of anti-peptide antibody for enrichment at the level of the peptide that has been enzymatically released from the protein biomarker of interest as part of the assay procedure (Fig. 1b). This approach is termed stable isotope standards and capture by anti-peptide antibodies (SISCAPA) [16]. Polyclonal and monoclonal anti-peptide antibodies can be used. Furthermore, bead-based [17] and column-based online flow formats [16, 18] are successfully used. Achievable sensitivity is in the pg/mL to low ng/mL range and above depending on the quality of the capture antibody and the sample volume used. One of the unique advantages of this workflow is the compatibility with harsh, denaturing conditions during sample preparation, for example as might be needed for extraction of protein biomarkers from tissues. To this end, a successful workflow might include tissue homogenization and extraction, protein precipitation followed by enzymatic digestion of the pellet prior to anti-peptide antibody-based enrichment. Harsh conditions during samples handling, for example using strong detergents, can be incompatible with protein IA methods irrespective whether the end-point detection is based on ligand binding or mass spectrometry.

Finally, protein and peptide IA methods can be combined (Fig. 1c) in a protein biomarker assay [11, 12]. Although the ***sequential protein and peptide IA*** approach is a more complex assay format, this configuration can deliver ultimate assay performance with respect to sensitivity and throughput. This assay format has been successfully employed for the routine analysis of thousands of clinical samples [14]. Which IA technique is selected depends on the availability of capture reagents, assay feasibility, available sample volume, required sensitivity, and other bioanalytical goals.

Liquid Chromatography Options

With the exception of some chromatography-free workflow developments, such as iMALDI [19], most quantitative protein biomarker MS assays require liquid chromatography (LC)-based separation of the analytes. Assays based on quantification of enzymatically released peptides mostly employ C18 reverse phase chromatography. Most LC configurations reported in the literature for quantitative protein assays use conventional, high flow rates typically at or above the mid-microlitre per minute range, with or without analyte trapping prior to analytical separation. These LC configurations are well tested and robust, mostly characterized by a short total cycle time and are easy to implement using standard equipment. In contrast, the

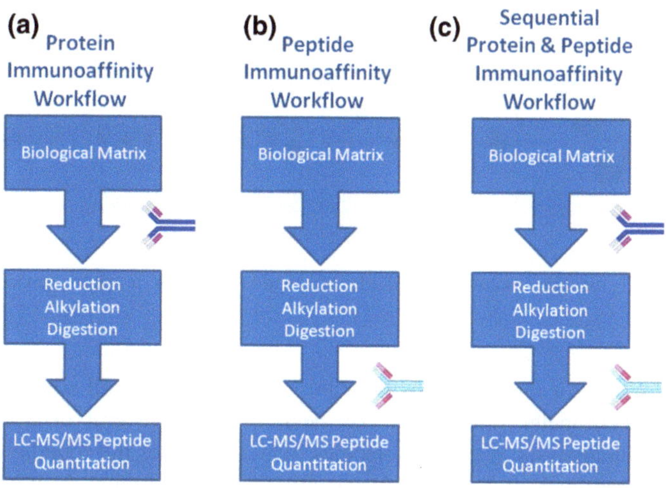

Fig. 1 Schematic of immunoaffinity workflows for LC-MS/MS quantification of protein biomarkers. **a** Immunoaffinity extraction of the protein biomarker from the sample using an anti-protein antibody prior to digestion and LC-MS/MS; **b** digestion of the sample into peptides followed by immunoaffinity extraction of the targeted peptide using an anti-peptide antibody prior to LC-MS/MS; **c** sequential protein immunoaffinity extraction followed by digestion and immunoaffinity extraction of the targeted peptide using an anti-peptide antibody

proteomics community has been using mostly nanoflow rates for reverse phase chromatography of complex digests, typically in the mid-nanoliter per minute range. This improves mass spectrometric sensitivity, which is inversely correlated with flow rate. Chromatography cycle times are typically longer (>1 h) in order to maximize separation for improved peptide identification or quantification. Advances in protein biomarker quantification workflows have illustrated the symbiosis between IA enrichment and nanoflow LC. Antibody directed analyte enrichment and therefore complexity reduction of the sample makes it possible to run short nanoflow gradients routinely for larger sample sets. Biomarker assays that employ such a workflow with total LC cycle times between 10 and 15 min have been published recently using either online anti-peptide antibody enrichment, offline anti-protein antibody enrichment, or both [11, 12, 14, 15, 18, 20]. Finally, capillary flow rates (low microlitre per minute range) are being used to bridge the sensitivity gap between methods that use high or nanoflow LC [21, 22].

Mass Spectrometry Techniques

Triple quadrupole mass spectrometers (Fig. 2) are most commonly used for quantification of surrogate, ***proteotypic peptides*** using the ***selected reaction monitoring*** (SRM) data acquisition mode. Selected fragment ions obtained from

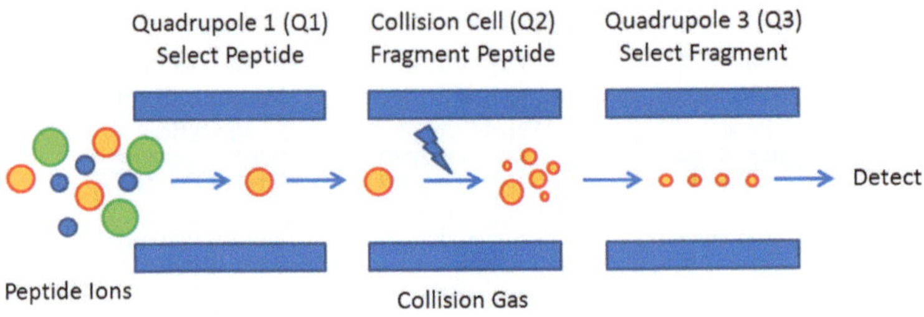

Fig. 2 Selected reaction monitoring (SRM) of peptide ions on a triple quadrupole mass spectrometer

predefined precursor ions via collision with gas in a collision cell are monitored by the mass spectrometer [23]. SRM has been used for several decades for the quantitative bioanalysis of small molecules, which provides the foundation for the recent advances made in LC-MS/MS quantification of proteins. In peptide SRM assays, typically 3–5 fragment ions per precursor are recorded, which are frequently those that are most abundant and free from interferences. These ion transitions are typically the most sequence informative, but that selection is not always needed particularly when paired with an immunoaffinity workflow that provides additional selectivity. A stable isotope labeled (SIL) standard is typically employed to coelute with the analyte of interest. This increases confidence in assay selectivity via correct assignment and quantification of the signal, especially from complex biological samples [24], including those from samples that have been enriched using immunoaffinity. Importantly, SIL peptide standards also mirror the expected intensity ratio of SRM transitions of the analyte which can be used to confirm correct signal assignment. Recommendations relating to the quantification, storage, and handling of peptide standards for mass spectrometry-based workflows have recently been published [25]. Finally, in addition to using recombinant protein calibrators where possible as well as endogenous and recombinant quality control samples, SIL peptide standards are utilized for MS response normalization as well as for normalizing parts of the sample preparation workflow (Table 1).

High-resolution (HR) MS instruments are increasingly explored for quantification of protein biomarkers via their surrogate peptides. Product ion scans on a QTOF mass spectrometer provide high measurement specificity. Akin to this approach, contemporary quantification methods on orbitrap MS instruments can use targeted higher energy collisional dissociation (tHCD) methods. Only precursor masses are preset and high-resolution and high mass accuracy allows simultaneous identification and quantification of multiple fragment ions from complex tandem mass spectra. Other HRMS quantification techniques rely on high measurement resolution and accuracy of the precursor peptide ions without fragmentation using selected ion monitoring (SIM) on both QTOF and orbitrap mass spectrometers [20]. A promising feature of HRMS is that in addition to the targeted quantification of the peptides of interest, the mass spectrometer can simultaneously acquire qualitative information from other components of the sample. Finally, quantification

Table 1 Normalization strategies based on heavy stable isotope labeled peptides and proteins for protein LC-MS/MS workflows

Normalization approach	Advantage	Disadvantage
Stable isotope labeled (SIL) peptide 	• Quickly synthesized • Inexpensive • Can normalize LC-MS	• Protein immunoaffinity and protein digestion not normalized
Extended sequence SIL peptide 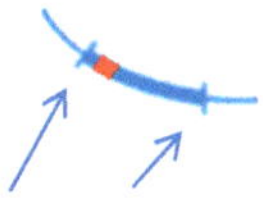 Enzymatic cleavage sites	• Quickly synthesized • Inexpensive • Includes digestion and LC-MS in normalization	• Protein immunoaffinity step not normalized
SIL protein 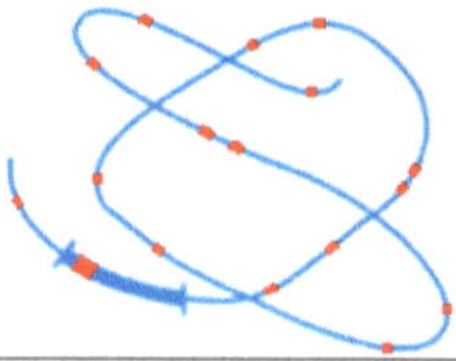	• Added at beginning of the assay to normalize entire workflow • Normalize multiple peptides	• Cost, time • Structural differences to endogenous analytes, such as folding, PTMs (may be problematic for protein immunoaffinity)

workflows that include protein immunoaffinity enrichment, but do not rely on digestion and instead analyze the intact protein by MS are beginning to be explored. This approach holds great potential, as workflows are simpler due to the absence of the digestion step. Furthermore, no structural or sequence information about the biomarker protein is lost when only one or a few peptides are monitored as surrogates. However, the currently achievable sensitivity is limited mostly due to multiple charging of the protein precursor during ionization, which makes this approach mostly suitable to quantification of biomarker proteins that are in higher abundance.

Building Translatable Assays

As candidate drug compounds and their targets and pathways progress through drug discovery stage gates toward investigations in clinical trials, the need for developing clinical biomarkers increases. Oftentimes, this necessitates de novo development of assays. Ideally, the same or similar bioanalytical method is used during the research and development continuum to facilitate better interpretation and translation of results between species and investigations. High measurement specificity and good sensitivity of protein biomarker LC-MS/MS are key drivers to implement the technology in translational research. However, additional opportunities exist to realize synergies for clinical assay development, for example when preclinical assays are also developed, optimizing the investment in bioanalytical resources. Specifically, a number of factors should be considered when developing protein biomarker LC-MS/MS assays that can be used across different species with no or only minor modifications. Assay translation between species can be facilitated by selecting ***proteotypic peptides*** that are conserved in the same protein across species, if available. This allows the use of the same MS detection method and the same SIL standard peptide. Furthermore, other reagents specific for the targeted peptides can then also be used in a cross-species assay, such as anti-peptide antibodies. In some cases, where the protein itself is highly homologous or conserved between the species of interest, cross-species binding of the antibody used for protein IA enrichment can be investigated as a desired reagent property. In a protein biomarker IA-LC-MS/MS assay, such a capture antibody may be preferentially chosen over other antibodies that do not cross-react. Examples of sequential protein and peptide IA-LC-MS/MS assays have been reported recently where all antibody reagents and SIL peptides were successfully employed across matrices from different species [11, 12, 15].

In the absence of conserved, cross-species sequences, the selection of peptides is guided by the analytical aim and the desired assay workflow. For example, if anti-peptide antibodies will be part of the assay, then peptides from the equivalent sequence region of the protein from different species can be selected that share a similar antigenic sequence as part of a ***proteotypic peptide***. A peptide immunogen sequence can then be carefully designed for generating anti-peptide antibodies, frequently in rabbits [16, 26], that are capable of binding the related ***proteotypic peptides*** from different species. The anti-peptide antibody reagent generated in such a way can be used in an IA-LC-MS/MS assay for enriching the relevant peptides from the different species. MS detection methods have to be adjusted accordingly.

Although quantitative protein biomarker LC-MS/MS has substantially evolved in recent years, it is still a fairly nascent technique. When developing a workflow based on this technique, perhaps using research grade instrumentation and methods, the technical implementation in a clinical setting needs to be carefully planned. Until recently, the operational complexity, particularly of immunoaffinity workflows prior to LC-MS/MS, has been perceived as a possible limitation to clinical implementation [27]. However, the technical challenges are being overcome by the use of standardized workflows, the implementation of liquid handling robotics, and robust mass spectrometry configurations [11, 12]. This led to recent examples of large-scale implementation of the quantitative LC-MS/MS technique in clinical protein biomarker studies [14]. Although a protein biomarker LC-MS/MS assay is typically developed under the fit-for-purpose paradigm [28], the assays can meet stringent acceptance criteria [11]. This technology is anticipated to mature further as additional precedence is generated and experience is gained in the field.

Chapter Summary

- LC-MS/MS, especially nanoflow LC-MS/MS, particularly when linked with immunoaffinity enrichment, has emerged as a viable bioanalytical technique for the quantitative measurement of protein biomarkers and therapeutic target proteins, impacting translational pharmacology.
- A key advantage of protein LC-MS/MS assay is the high measurement specificity that can be achieved.
- Immunoaffinity enrichment techniques using anti-protein or anti-peptide antibodies, or both, extend LC-MS/MS assay sensitivity for protein biomarkers into the pg/mL range.
- Assay translation between species can be facilitated by selecting proteotypic peptides that are conserved in the same protein across species, if available, to allow the same MS detection method, the same SIL standard peptide and the same anti-peptide antibody can be used.
- Practical challenges to routine implementation in clinical assays are being overcome by the use of standardized workflows, liquid handling robotics, and robust LC-MS/MS configurations.

References

1. Rifai N, Gillette MA, Carr SA (2006) Protein biomarker discovery and validation: the long and uncertain path to clinical utility. Nat Biotechnol 24(8):971–983
2. Mallick P, Kuster B (2010) Proteomics: a pragmatic perspective. Nat Biotechnol 28(7): 695–709
3. Coombs KM (2011) Quantitative proteomics of complex mixtures. Expert Rev Proteomics 8(5):659–677
4. Egertson JD, Kuehn A, Merrihew GE et al (2013) Multiplexed MS/MS for improved data-independent acquisition. Nat Methods 10(8):744–746
5. Fernandez Ocaña M, Neubert H, Przyborowska A et al (2004) BSE control: detection of gelatine-derived peptides in animal feed by mass spectrometry. Analyst 129(2):111–115
6. Neubert H, Grace C, Rumpel K, James I (2008) Assessing immunogenicity in the presence of excess protein therapeutic using immunoprecipitation and quantitative mass spectrometry. Anal Chem 80(18):6907–6914
7. Neubert H, James I (2009) Online capillary weak cation exchange enrichment hyphenated to nanospray mass spectrometry for quantitation of a basic pegvisomant derived peptide. J Chromatogr A 1216(33):6151–6154
8. Anderson NL, Anderson NG (2002) The human plasma proteome: history, character, and diagnostic prospects. Mol Cell Proteomics 1(11):845–867
9. Jian W, Edom RW, Wang D, Weng N, Zhang S (2013) Relative quantitation of glycoisoforms of intact apolipoprotein C3 in human plasma by liquid chromatography-high-resolution mass spectrometry. Anal Chem 85(5):2867–2874
10. Ackermann BL, Berna MJ (2007) Coupling immunoaffinity techniques with MS for quantitative analysis of low-abundance protein biomarkers. Expert Rev Proteomics 4(2):175–186
11. Neubert H, Muirhead D, Kabir M, Grace C, Cleton A, Arends R (2013) Sequential protein and peptide immunoaffinity capture for mass spectrometry-based quantification of total human beta-nerve growth factor. Anal Chem 85(3):1719–1726

12. Palandra J, Finelli A, Zhu M, Masferrer J, Neubert H (2013) Highly specific and sensitive measurements of human and monkey interleukin 21 using sequential protein and tryptic peptide immunoaffinity LC-MS/MS. Anal Chem 85(11):5522–5529
13. Fernández Ocaña M, Neubert H (2010) An immunoaffinity liquid chromatography-tandem mass spectrometry assay for the quantitation of matrix metalloproteinase 9 in mouse serum. Anal Biochem 399(2):202–210
14. Schultz GA, Mccardle K, Neubert H (2016) Large-scale implementation of sequential protein and peptide immunoaffinity enrichment LC/nanoLC–MS/MS for human β-nerve growth factor. Bioanalysis 8(8):753–764
15. Palandra J, Quazi A, Fitz L, Rong H, Morris C, Neubert H (2016) Quantitative measurements of GDF-8 using Immunoaffinity LC-MS/MS. Proteomics Clin Appl 10(5):597–604. doi:10.1002/prca.201500112n/a-n/a
16. Anderson NL, Anderson NG, Haines LR, Hardie DB, Olafson RW, Pearson TW (2004) Mass spectrometric quantitation of peptides and proteins using Stable Isotope Standards and Capture by Anti-Peptide Antibodies (SISCAPA). J Proteome Res 3(2):235–244
17. Whiteaker JR, Zhao L, Anderson L, Paulovich AG (2010) An automated and multiplexed method for high throughput peptide immunoaffinity enrichment and multiple reaction monitoring mass spectrometry-based quantification of protein biomarkers. Mol Cell Proteomics 9(1):184–196
18. Neubert H, Gale J, Muirhead D (2010) Online high-flow peptide immunoaffinity enrichment and nanoflow LC-MS/MS: assay development for total salivary pepsin/pepsinogen. Clin Chem 56(9):1413–1423
19. Shah B, Reid J, Kuzyk M, Parker C, Borchers C (2013) Developing an iMALDI Method. Methods Mol Biol 1023:97–120
20. Fan Y-Y, Neubert H (2016) Quantitative Analysis of Human Neonatal Fc Receptor (FcRn) tissue expression in transgenic mice by online peptide immuno-affinity LC-HRMS. Anal Chem. doi:10.1021/acs.analchem.5b03900
21. Wang H, Bennett P (2013) Performance assessment of microflow LC combined with high-resolution MS in bioanalysis. Bioanalysis 5(10):1249–1267
22. Lee AYH, Chappell DL, Bak MJ et al (2016) Multiplexed quantification of proglucagon-derived peptides by immunoaffinity enrichment and tandem mass spectrometry after a meal tolerance test. Clin Chem 62(1):227–235
23. Picotti P, Aebersold R (2012) Selected reaction monitoring-based proteomics: workflows, potential, pitfalls and future directions. Nat. Meth. 9(6):555–566
24. Sherman J, Mckay MJ, Ashman K, Molloy MP (2009) How specific is my SRM? The issue of precursor and product ion redundancy. Proteomics 9(5):1120–1123
25. Hoofnagle AN, Whiteaker JR, Carr SA et al (2016) Recommendations for the generation, quantification, storage, and handling of peptides used for mass spectrometry-based assays. Clin Chem 62(1):48–69
26. Pope ME, Soste MV, Eyford BA, Anderson NL, Pearson TW (2009) Anti-peptide antibody screening: selection of high affinity monoclonal reagents by a refined surface plasmon resonance technique. J Immunol Methods 341(1–2):86–96
27. Ackermann BL (2012) Understanding the role of immunoaffinity-based mass spectrometry methods for clinical applications. Clin Chem 58(12):1620–1622
28. Lee J, Devanarayan V, Barrett Y et al (2006) Fit-for-purpose method development and validation for successful biomarker measurement. Pharm Res 23(2):312–328

LC/MS Methods for Small Molecule Biomarkers

Michael P. Sullivan

Abstract Small molecule biomarkers provide insight into metabolic pathways related to therapeutic treatment and can help explain the mechanism of action during drug development. LC-MS-MS method approaches parallel those for drug pharmacokinetic analyses; the optimization of detection, chromatographic system, and extraction are similarly important to achieving good quantitation. Endogenous presence of small molecule biomarkers presents a unique challenge for developing and validating an LC-MS-MS assay and several strategies are available to address it. Additional challenges for small molecule biomarker assays are sensitivity, ion suppression/matrix effects, and assay selectivity. Innovative solutions for overcoming these hurdles are discussed in this chapter.

Keywords Mass spectrometry · LC-MS-MS · Chromatography · HPLC · Extraction · Endogenous · Surrogate · Matrix · Sensitivity · Selectivity

Introduction

Biomarkers come in all shapes and sizes. A lot of emphasis has been placed on proteins and genetic processes, but small molecules can provide information not found with larger molecules with regard to metabolic pathways, pathology, causes, and potential treatment. In this chapter, the role of small molecule biomarkers will be discussed briefly, followed by what comprises a liquid chromatography mass spectrometric (LC-MS-MS) assay. The challenges inherent to LC-MS-MS biomarker analyses and some examples of current small molecule biomarker assays will also be covered.

M.P. Sullivan (✉)
Department of Bioanalytical Sciences, Worldwide Clinical Trials,
8609 Cross Park Drive, Austin, TX 78754, USA
e-mail: michael.sullivan@worldwide.com

R. Weiner and M. Kelley (eds.), *Translating Molecular Biomarkers into Clinical Assays*, AAPS Advances in the Pharmaceutical Sciences Series 21,
DOI 10.1007/978-3-319-40793-7_10

Drug Development—Efficacy

Metabolomics provides a vast number of potential biomarkers for measuring how well a pharmaceutical compound is working. Energy substrates and metabolites are monitored for changes in diabetics during treatment. Cholesterol and lipid levels provide feedback for the progress cardiac patients are making on lipid lowering medications. For new drugs under testing, these same markers are valuable tools for establishing how they are producing their effect as well as allowing a comparison with established treatment lines.

Drug Development—Safety

Biomarkers related to drug safety is a relatively new area in drug development with only a few recognized and validated protein compounds targeting renal and liver function. Small molecule safety biomarkers are of interest, but not well investigated. Any compound that may be linked to morbidity and mortality are candidates to be used. Blood lipids are an example of potential safety indicators for long-term effects related to cardiovascular disease.

Advantages of LC-MS-MS Measurements

Sensitivity

Tandem mass spectrometry brings extreme sensitivity to detection of analytes when coupled to HPLC applications. Background noise is greatly reduced by the selectivity of single ion mass monitoring (SRM or MRM mode), elevating the signal-to-noise ratio over less selective detectors.

Selectivity/Specificity

While signal-to-noise is increased, interferences are reduced through the filtering of all but the target mass ranges of interest. Comparisons of HPLC-UV and LC-MS-MS demonstrate the much shorter chromatographic run times of LC-MS-MS techniques [1].

Development of LC-MS-MS Assays

Mass Spec Optimization

Just as in the bioanalysis of an exogenous compound, endogenous biomarker analysis by LC-MS-MS first begins with the optimization of the mass spectrometer in order to detect the compound of interest. Mass selection (of precursor and product ions) and settings optimizations are the tasks involved in order to give the best response for a given amount of material. This can be accomplished through flow injection or constant infusion of the compound while monitoring in a scanning mode. Lens settings are ramped in order to find the optimal settings that promote the greatest signal for the masses found.

Liquid Chromatography (LC) Optimization

Of course the optimizing of the LC system provides an important aspect of the selectivity of the overall method by separating potential interferences from the analyte of interest through the orderly retention and elution from the LC column.

However, the selection of the HPLC parameters used in tandem with mass spectrometric detection can have a significant effect on sensitivity as well. Specifically, mobile phase composition and flow rate can either improve or inhibit the response of the analyte, depending on what additives and parameters are used. Generally, mobile phase additives that promote an easier evaporation (desolvation) when the liquid stream is nebulized in the mass spectrometer source, improves the transmission of ions to the mass spectrometer analyzer.

The decisions regarding the LC system for a given method will have an extensive effect on the sensitivity. Compounds that are very polar that retain poorly on a reverse phase column (C18) typically run at a mobile phase composition of 10 % organic solvent (methanol, acetonitrile) and 90 % aqueous solvent (water). The sensitivity of these compounds suffers under these conditions as the nebulized droplets from the LC flow struggle to evaporate and reduce in size. However, these same compounds are more retained on the polar ligands of normal phase or HILIC columns, and the mobile phase compositions at elution tend to be in the 80–90 % organic solvent range. The higher organic solvent composition makes it easier for desolvation in the mass spec source and improves sensitivity.

Some compounds show signal sensitivity to acid and base additives in the mobile phase as well. This may be due to ion pairing with the analyte, competitive protonation (in positive ion mode), or surface tension effects of the nebulized droplets in the mass spectrometer source. Regardless of the etiology, it is worthwhile to screen for these responses to mobile phase composition if sensitivity of the biomarker method is going to be one of the major challenges of development.

Sample Preparation Techniques

The simplest preparation technique is the direct injection of sample, presumably plasma or urine, into the LC solvent stream where it is cleaned up on-line through a trapping column or in-line solid phase extraction (SPE) [2]. This type of technique produces challenges in preserving the stability of the thawed, unaltered sample as it waits for processing in the sequence queue. However, it is the simplest form of manual preparation, as the sample is placed directly into the system where aliquot aspiration and processing are completely automated.

Most forms of sample preparation fall under the categories of sample dilution or protein precipitation (plasma samples), liquid–liquid (two-phase) extraction, or solid phase extraction (SPE).

Sample dilution of non-protein containing samples is a process where a quantitative aliquot of sample (urine) is combined with an internal standard and further diluted with a reagent compatible with injection. The advantages are that there is very little manipulation leading to little going wrong in the extract preparation. This technique is used for assays that have relatively high quantitative ranges where the detector signal can afford to be diluted.

Key Term 1
Internal Standard: A related compound added at predefined stage of the sample preparation process to provide a method for normalizing variations in analyte recovery, autosampler injection, and instrument response. The same amount is added to all samples within a batch and the ratio of analyte response divided by the internal standard response is used for regression analysis and quantitation of samples.

For protein containing samples (plasma, tissue homogenate), there is the potential for protein to precipitate as the diluted sample is introduced to the solvent stream of the LC system. This causes the HPLC column to clog and give high back pressure to the pumping system and can cause secondary chromatographic artifacts in the form of peak shape aberrations. Therefore, protein precipitation of these samples off-line, prior to injection, is performed as part of the extract preparation. The tertiary structure of the protein is forced to rearrange and fold in on itself by using a reagent, leading to insolubility and precipitation. The supernatant of the precipitated sample can be injected directly onto the LC system, or it can be evaporated to dryness and reconstituted in a more appropriate solvent. The reagents preferred for LC-MS applications are aqueous soluble solvents such as acetonitrile, methanol, and acetone. Strong acids such as trichloroacetic acid, perchloric acid, and trifluoroacetic acid are usually avoided as precipitating reagents due to their effects on ionization or their poor volatility in the instrument source.

Liquid–liquid extraction utilizes an immiscible organic solvent to form two layers of liquid separated by an interface. The analyte of interest starts out in the aqueous layer of the sample aliquot and diffuses across the interface until equilibrium is reached. The organic solvent is then removed and evaporated to dryness before reconstituting for injection. Numerous buffers can be added to the sample to encourage a greater partitioning to the organic solvent. Different forms of mixing are used to accelerate the equilibration process. This type of extraction produces a very clean final extract, as a lot of interferences, proteins, and phospholipids are left behind with the aqueous layer. It also allows for a fast and easy process for concentrating the sample when additional sensitivity is required.

Solid phase extraction (SPE) is a technique where a mixed liquid sample is applied to a solid resinous sorbent contained in either a plastic cartridge or the well of a 96-well extraction plate. The sorbent may contain one of several types of chemical ligands that function to interact and retain compounds with certain functional groups or chemical properties. An SPE reverse phase sorbent will have affinity for lipophilic components of the sample. An SPE strong cation exchange sorbent will attract and retain compounds with basic functional groups. Once the analyte is adsorbed to the solid phase of the cartridge/well, other components of the sample can be rinsed away with selected wash reagents. When it is time to recover the analyte from the sorbent, an eluting solution is passed through the cartridge/well and the adhered compound releases and collects in a test tube or 96-well block. This eluate is often evaporated to dryness and reconstituted before injection, but on certain occasions it can be injected without evaporation. This technique can produce a very clean extract and is very effective for more polar compounds, which can otherwise be difficult to extract without significant loss of recovery.

Challenges of Biomarker Analysis by LC-MS-MS

Most of the challenges with small molecule biomarker analysis relate to their being endogenous to the control matrix that would otherwise be used for calibration and quality control samples of targeted concentration. The underlying native concentration of the compound makes it difficult to determine the accuracy of any preparations used for quantitative assessments.

Calibration samples that are fortified to a target concentration will be biased by the endogenous level present in the lot of control matrix used. This can be subtracted from each sample to give an adjusted regression of the calibration curve, but it must be recalculated with each change in control matrix lot, and can often raise a question about the true quantitation/detection limit of the analysis if the background contribution is high compared to the lowest target concentration. The run–to-run variability related to background subtraction is the reason other alternatives are just as popular for addressing the quantitation of endogenous compounds.

To avoid background subtraction of endogenous compounds, calibration samples need to be prepared in a comparable matrix (to the experimental samples) which is free of the analyte being measured. This can be achieved in a few different ways. Matrix lots can be screened for suitably low levels (below detection or quantitation) of the endogenous compound. The control matrix can be altered to remove or destroy the endogenous compound before using for calibration sample preparation. An alternative species can be used as a control matrix source if the presence of the analyte is species dependent. Also, a proxy matrix can be prepared that is free of the endogenous compound [3, 4]. In all of these cases where a substitute matrix is used, testing to ensure comparable accuracy is achieved between the authentic matrix and the surrogate by way of standard addition to the authentic matrix and measuring the accuracy of the addition [3, 5]. Parallelism of calibration curves prepared in the authentic and surrogate matrix provides similar support for the alternate matrix [6].

Determining the quantitative range of a small molecule biomarker assay can sometimes be difficult compared to selecting the upper and lower limits of a pharmacokinetic assay. When an assay is used for drug development purposes, more is needed than simply having a positive or negative response from the biomarker. A quantitative measure is typically needed in order to be able to apply statistical analyses when looking at effectiveness of a treatment. This requires measuring baseline levels under normal conditions as well as under conditions of upregulated or downregulated influence.

Where an inhibited response is expected under treatment, sensitivity for the biomarker then becomes an issue. Measures to optimize sensitivity, such as low flow rates, high organic composition of mobile phase, increased sampling size, or in extreme cases derivatization of the analyte may need to be employed. Where an enhanced response is expected under treatment, determining the highest responses becomes the challenge in order to set the upper limit of the assay. With larger dynamic calibration ranges comes issues with linearity of the detector and contamination/carryover from automation equipment and injection instrumentation.

Key Term 2
Derivatization: A process used in analytical methods to change the molecular structure of a compound with a reagent that reacts with one or more functional groups on the analyte. Examples of derivatized products are esters, amides, oximes, and hydrazones. This process can help change the characteristics of the compound for the purpose of analysis. For example, a derivative may be employed in order to improve the chromatographic behavior of a compound.

Current Small Molecule LC/MS Methods

Existing methods for small molecule biomarker analysis by LC-MS are numerous, and the details of these methods are not always available through publication. Table 1 includes a listing of methods found across several therapeutic areas. Additionally, selected methods are discussed with the details of how the inherent difficulties of the compounds were overcome.

Table 1 Other small molecule biomarkers by LC-MS

Disease area	Disease	Compound	Matrix	Reference
Neurology	PD, PAF, Depression	DOPAC, DHPG	CSF	[23]
		Serotonin, 5-HIAA	CSF	[24–26]
Oncology	Melanoma	5-S-Cysteinlydopa	Plasma	[27]
	Intestinal carcinoid tumor	5-HIAA	Urine	[2]
Metabolism	Fatty acid amide hydrolase inhibition	Ethanolamides	Plasma	[28]
	In vivo CYP3A4/5 activity	4β-Hydroxycholesterol/ cholesterol	Plasma	[3]
Cardiology	Atherosclerosis	Eicosanoids	Plasma	[4]
Environmental	Toluene exposure	Hippuric acid	Urine	[29]
Pulmonary Disease	COPD	Desmosines	Plasma, urine	[30]

Abbreviations DOPAC = dihydroxyphenylacetic acid, DHPG = dihydroxyphenylglycol, 5-HIAA = 5-hydroxyindoleacetic acid, PD = Parkinson's Disease, PAF = pure autonomic failure, COPD = chronic obstructive pulmonary disease

Neurotransmitters

Monoamine neurotransmitters (Fig. 1) have been used as markers of neurological health and function. Biologically, they are easily synthesized from simple amino acid reserves to yield the active signal transmitters. These active compounds include norepinephrine, epinephrine, dopamine, and serotonin, among others. Measurements of their metabolites are also very helpful in determining mechanism of action of certain therapeutic agents (reuptake inhibitors) and serving as markers themselves in various pathological states (DOPAC and DHPG in Parkinson's Disease).

Traditionally, neurotransmitter analysis has been accomplished through the application of HPLC with electrochemical detection, due to the majority of analytes having very little absorbance activity for UV detection and their sensitivity to oxidation. However, these methods often suffered from poor sensitivity, interferences from complex samples, and extremely long analysis times [7]. More recent applications using LC-MS-MS have improved these areas of their analyses.

From an analytical perspective, neurotransmitters are very small and polar in nature. Their elution profiles on reverse phase HPLC shows them eluting early, under very weak chromatographic conditions. Extraction methods used for sample preparations for HPLC applications utilized alumina SPE, taking advantage of the polar diol functional groups of the catechol structure [8].

These characteristics make the analysis of this class of compounds difficult. For an LC-MS-MS application, a low molecular weight produces a high noise level, reducing sensitivity, and selectivity. Poor retention on reverse phase LC systems produces a poor desolvation condition, further limiting sensitivity.

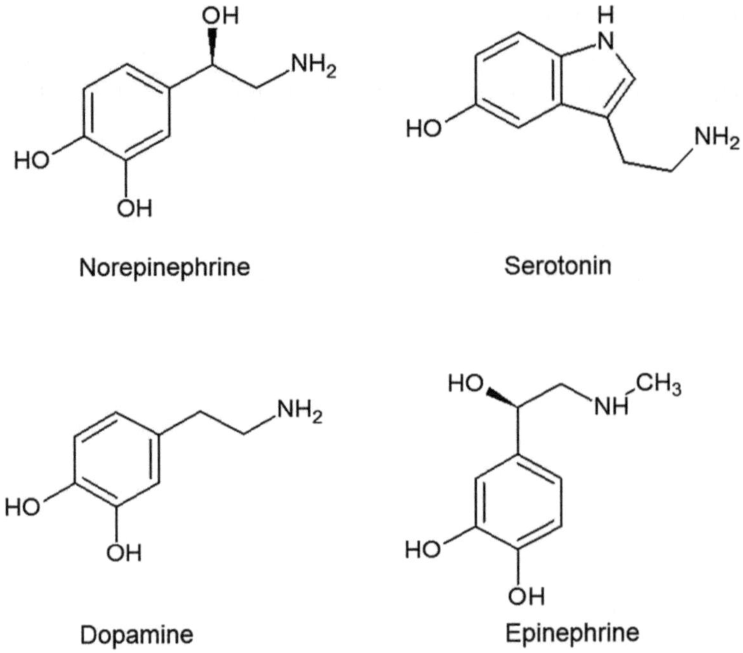

Fig. 1 Chemical structures of neurotransmitters (Norepinephrine, Serotonin, Dopamine, Epinephrine)

By using derivatization with dansyl chloride, issues with sensitivity can be improved [9]. The dansyl derivative leaves a larger molecule for chromatographic and mass spectrometric analyses. With better retention and lower noise from the higher ion transitions, sensitivity and selectivity are improved. In addition, several acid metabolites can also be analyzed under the same conditions that would otherwise require the opposite ionization polarity (negative ion electrospray). A similar solution to LC-MS-MS chromatographic retention and sensitivity through ethylation of epinephrine and norepinephrine after alumina extraction has been used. This allowed for a sensitive assay achieving detection limits to the low pg/mL level in plasma [10].

Derivatization of low molecular weight compounds or poorly ionized compounds can provide a powerful leverage to improve sensitivity by LC-MS-MS.

Vitamin D

Vitamin D analysis is a recent addition to LC-MS-MS techniques. Vitamin D levels in plasma are measured to assess the status of the individual, where low levels are associated with heart disease, diabetes, cancer, autoimmune disorders, and of course bone growth disorders [11].

Vitamin D originates from both endogenous (vitamin D_3) and dietary (vitamin D_2) sources. Both forms are quickly metabolized in the liver to 25-hydroxyvitamin D forms (25-(OH)D_3 and 25-(OH)D_2) which circulate with relatively long half-lives until they are converted to their active 1,25-dihydroxyvitamin D forms in the kidney. The active dihydroxy forms have a very transient existence and are present at extremely low levels. For these reasons, the concentrations of plasma 25-(OH)D_2 and 25-(OH)D_3 are considered the reference biomarkers for vitamin D nutritional status [11].

LC-MS-MS analysis of vitamin D provides all the expected advantages the hyphenated technique brings to other assays, which include improved sensitivity and selectivity, reduced processing and run times, and the ability to measure multiple analytes in a single analysis. However, limitations of the technique are demonstrated in the case of vitamin D. Common, nonspecific water loss product ion fragments make interferences from analogous compounds a potential problem. Also, an epimer of 25-(OH)D_3 is known to interfere on LC-MS-MS assays, while there is no such interference with immunoassays for vitamin D. These challenges are overcome with careful selection of mass spec settings and chromatographic conditions [11, 12].

For the analysis of 25-(OH)D_2 and 25-(OH)D_3 (Fig. 2), assays utilizing many different variables and setups have been successfully validated and used for clinical and drug development purposes. Electrospray as well as atmospheric pressure chemical ionization (APcI) sources have been used. Electrospray is known to be more susceptible to ion suppression events during analysis, and APcI used with vitamin D analysis has been shown to provide better precision of results presumably due to less suppression effects [12].

Product ion fragments using water loss have been commonly used for high signal response, but they have been subject to interferences not experienced with more substantial structural fragmentations. LC-MS-MS can easily distinguish between the precursor ions of 25-(OH)D_2 and 25-(OH)D_3 (413 and 401, respectively). However other hydrophobic lipids or structurally similar sterols have the potential to share these masses and a water loss transition is not unusual for this class of compounds. Listed transitions for 25-(OH)D_2 that have been used are 413 > 395, 413 > 377, 413 > 355, 413 > 337, 413 > 83, and for 25-(OH)D_3 are 401 > 365, 401 > 257, 401 > 159. During APcI ionization, the use of precursor masses with nominal water loss (395 for 25-(OH)D_2 and 383 for 25-(OH)D_3) is also common [12–14].

Sample preparation techniques used for vitamin D analysis include protein precipitation, liquid–liquid extraction, and solid phase extraction. The simpler extraction methods give shorter sample processing times, but often require more involved chromatographic separation and run time. One example of a method utilizing acetonitrile precipitation and filtering of the extract still required an LC system incorporating a switching valve and trapping column to effectively remove phospholipid interferences [13]. Liquid–liquid extraction techniques have been very effective for eliminating phospholipid interferences and allowing for shorter LC analysis times [12–14].

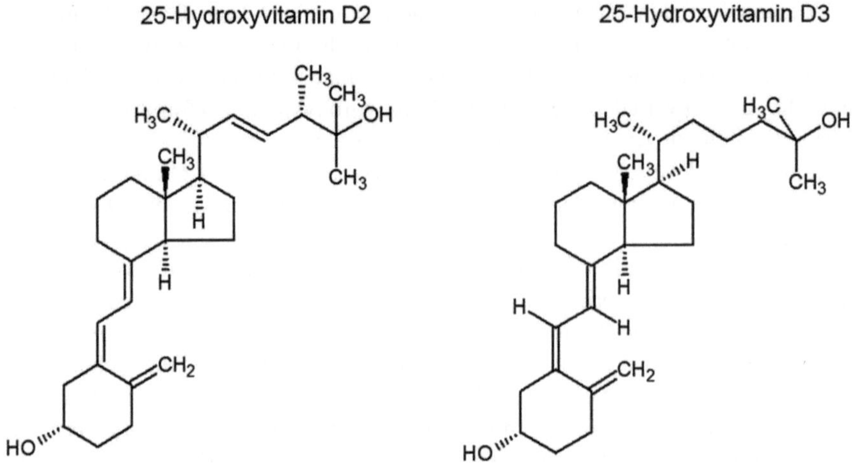

Fig. 2 Chemical structures of Vitamin D analytes (25-Hydroxyvitamin D2, 25-Hydroxyvitamin D3)

Steroids

Endogenous steroids are generally very lipophilic compounds that tend to ionize poorly, making them difficult to measure by LC-MS-MS, and yet, it is becoming the method of choice for steroid analysis. Advantages in analysis time, specificity, and performance at low concentrations make LC-MS-MS preferred over immunoassays.

The main challenges in steroid analysis by LC-MS-MS are poor ionization for sensitivity and susceptibility to interference from isobaric compounds. Recent methods have brought innovative approaches to deal with these problems.

In the analysis of estrone and estradiol (Fig. 3), very low detection limits are needed to quantify plasma samples in pre-pubescent females, men, and post-menopausal women. Methods that measure as low as 1 pg/mL are needed to detect even normal basal levels for the purpose of clinical analysis or drug development. Although both analytes are lipophilic, their chemical structures contain a phenolic hydroxyl group that can ionize in negative ion mode. This is a weakly acidic functional group, but the addition of ammonium fluoride in the mobile phase has been shown to improve the sensitivity of monitoring the native estrogen compounds presumably through improved ionization in negative ion mode [15, 16].

Liquid–liquid extraction appears to be the sample preparation of choice to give as clean an extract as possible [15]. Serial extraction is also sometimes employed to achieve improved recovery [16]. With such a low detection limit, further clean up with a trapping column prior to separation on the analytical LC column has also been used [16]. Dansyl derivatization of the phenolic hydroxyl group after extraction has been applied to estrone and estradiol in order to give an easily ionizable group to the structure [17]. This provides a way to improve the detection

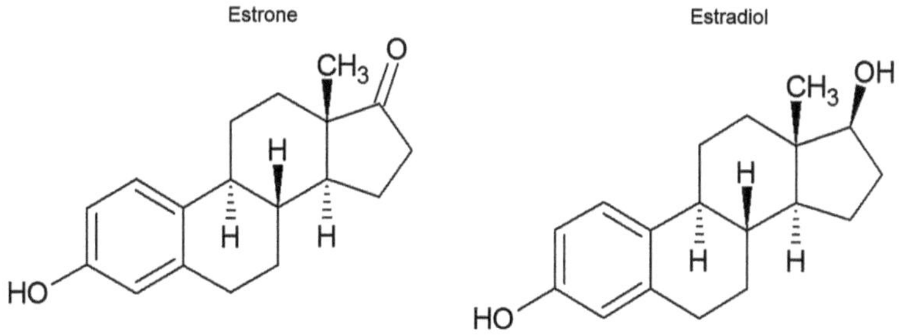

Fig. 3 Chemical structures of estrone and estradiol (Estrone, Estradiol)

sensitivity, but all applications seem to require a very selective chromatographic system in order to ensure required resolution between estrone and estradiol, as well as other potential isobaric interferences [15–17]. Other endogenous steroid methods follow similar schemes; extremely low detection limit, lipophilic extraction and concentration, derivatization (dansyl chloride, hydroxylamine), and a very selective chromatographic system often utilizing UHPLC [15, 18] to avoid isobaric interferences.

Key Term 3
UHPLC (Ultra High Pressure/Performance Liquid Chromatography): A chromatographic system similar to HPLC, but uses a smaller particle packing in the column (usually <2 microns). The smaller particles are closely packed and cause a high back pressure from the column, but peaks are sharper and resolution of peaks is obtained more rapidly.

Nicotine/Cotinine

Environmental tobacco smoke exposure is a concern from a clinical perspective due to its general effects on health (risk factor for cardiovascular diseases, cancers, and deaths) and in particular its impact on organ transplant. Smoking cessation efforts are considered an integral part of optimal solid organ transplant for tissue donors and recipient candidates. As an alternative to self-reporting, nicotine status is proposed as a biomarker for tobacco smoke exposure. Nicotine is one of the major components of tobacco which is quickly absorbed when inhaled and distributed in the blood. It is metabolized within a couple of hours to cotinine in the liver. Cotinine has a relatively long half-life in circulation (approximately 6–22 h), which gives it one advantage as a biomarker over short-lived candidates [19, 20].

Nicotine and cotinine (Fig. 4) are very small molecules, which makes their detection by LC-MS-MS as challenging as the neurotransmitters discussed earlier.

Fig. 4 Chemical structures of nicotine and cotinine (Nicotine, Cotinine)

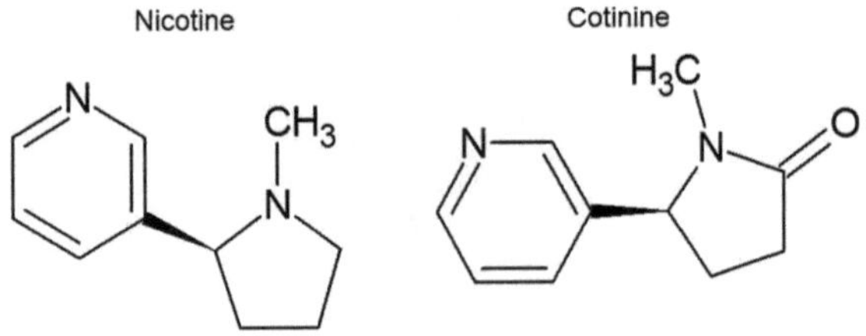

Having low molecular weights allows for many interferences from the low end of the mass scale and provides background noise that limit sensitivity. Fortunately, the detection limits for the purpose of tobacco exposure are relatively high (1–2 ng/mL) which allows for the monitoring of the native underivatized structures. Common transitions used are 163 > 130 for nicotine and 177 > 80 for cotinine using electrospray positive ion mode. Chromatographic conditions are oriented toward reverse phase on long (>100 mm) HPLC columns under gradient elution. Sensitivity of the detector response to mobile phase components is not an issue, as mobile phase compositions tend toward traditional preparations; acetonitrile, methanol, and water, with and without formic acid and acetate salts [19–21].

Extractions for combined assays of nicotine and cotinine usually are performed using both ion exchange and reverse phase SPE. Due to the sensitivity requirements, the extracts are concentrated through evaporation prior to injection. Nicotine can be volatile compared to other small molecules, so in some cases an acid keeper (hydrochloric acid) is added before the drying step [19–21].

One of the more challenging aspects of the analysis of nicotine and its metabolites is finding clean matrix for use with calibration samples. Conceivably, identifying nonsmoking individuals should be sufficient to harvest nicotine-free plasma, but the presence of second-hand exposure to tobacco smoke, as well as potential dietary sources of nicotine are enough to produce responses that bias the lower level calibrators. For some, the use of pre-screened matrix from commercial sources is sufficient for use, but others found the need to scrub serum or purchase a nicotine-free synthetic matrix to have acceptably clean control blank samples [19–21].

When comparisons of different nicotine methods by LC-MS-MS have been made through analysis of common samples, good reproducibility of results has been reported [19]. This suggests the analysis of nicotine and cotinine is very reliable, even across different sample preparation and chromatographic conditions, once the sensitivity and endogenous presence challenges are under control. Correlation between nicotine response and cotinine response (corrected for sampling time) after dosing (nicotine patch) has been demonstrated to be good, confirming the utility of cotinine analysis as a nicotine status indicator [21].

The Future of Small Molecule Biomarker Analysis by LC/MS

As small molecule biomarkers become a greater part of the drug development process, the future holds several areas of expansion. One is the standardization of validation approaches for establishing a biomarker. In a regulated environment, specific recommendations are going to be made on how to approach the preparation of calibration curves for endogenous compounds and how to evaluate method validation parameters such as selectivity, limit of quantitation, and matrix effects. In addition, instrumentation capabilities are bound to improve. As technology improves in LC-MS-MS detectors, quantitation of very low level metabolites will be possible. This will open up the scope of possible biomarker candidates available for evaluation, expanding an already plentiful metabolome database [22]. Regardless of those advances, exploring the existing contents of the blood, CSF, and urine metabolomes brings more small molecule biomarker methods to light.

Key Term 4
Matrix Effect: The influence on analytical response (of the detector) by the presence of plasma, urine, or CSF (matrix) components in an extracted sample, as compared to a sample containing only solvent (no matrix). The presence of matrix can effect LC-MS ionization and it can also influence extraction recovery. This term is used to help explain inconsistencies of chromatographic peak responses from different samples during analysis.

Chapter Summary

- Small molecule biomarkers provide insight into metabolic pathways related to therapeutic treatment and can help explain the mechanism of action during drug development.
- LC-MS-MS method approaches parallel those for drug pharmacokinetic analyses; optimizations of detector, chromatographic system, and extraction are similar.
- Endogenous presence of small molecule biomarkers present a unique challenge for developing and validating an LC-MS-MS biomarker assay compared to excipient compounds.
- Other challenges and innovative solutions for small molecule biomarker assays:
  - *Sensitivity*—derivatization and optimized mobile phase composition are strategies that can improve signal response in the mass spectrometer.
  - *Ion suppression*—use of APcI can reduce ion suppression effects seen in electrospray ionization.

Selectivity—use of alternate phases (ion exchange, HILIC) besides reverse phase chromatography provides retention and separation based on different characteristics. UHPLC can increase throughput by resolving critical peak pairs in short time frames.

References

1. Georgita C, Sora I, Albu F, Monciu CM (2010) Comparison of a LC/MS method with a LC/UV method for the determination of metformin in plasma samples. Farmacia 58(2): 158–169
2. Perry H, Keevil B (2008) Online extraction of 5-hydroxyindole acetic acid from urine for analysis by liquid chromatography-tandem mass spectrometry. Ann Clin Biochem 45(2): 149–152
3. Xu Y, Yuan Y, Smith L, Edom R, Weng N, Mamidi R, Silva J, Evans DC, Lim HK (2013) LC-ESI-MS/MS quantification of 4β-hydroxycholesterol and cholesterol in plasma samples of limited volume. J Pharm Biomed Anal 85:145–154
4. Rago B, Fu C (2013) Development of a high-throughput ultra performance liquid chromatography-mass spectrometry assay to profile 18 eicosanoids as exploratory biomarkers for atherosclerotic diseases. J Chromatogr B 936:25–32
5. Wilson SF, James CA, Zhu X, Davis MT, Rose MJ (2011) Development of a method for the determination of glycine in human cerebrospinal fluid using pre-column derivatization and LC-MS/MS. J Pharm Biomed Anal 56:315–323
6. Houghton R, Pita CH, Ward I, Macarthur R (2009) Generic approach to validation of small-molecule LC-MS/MS biomarker assays. Bioanalysis 1(8):1365–1374
7. de Jong WHA, de Vries EGE, Kema IP (2011) Current status and future developments of LC-MS/MS in clinical chemistry for quantification of biogenic amines. Clin Biochem 44: 95–103
8. Holmes C, Eisenhofer G, Goldstein DS (1994) Improved assay for plasma dihydroxyphenylacetic acid and other catechols using high-performance liquid chromatography with electrochemical detection. J Chromatogr B Biomed Sci Appl 653(2):131–138
9. Cai HL, Zhu RH, Li HD (2010) Determination of dansylated monoamine and amino acid neurotransmitters and their metabolites in human plasma by liquid chromatography-electrospray ionization tandem mass spectrometry. Anal Biochem 396:103–111
10. Zhang G, Zhan Y, Ji C, McDonald T, Walton J, Groeber EA, Steenwyk RC, Lin Z (2012) Ultrasensitive measurement of endogenous epinephrine and norepinephrine in human plasma by semi-automated SPE-LC-MS/MS. J Chromatogr B 895–896:186–190
11. El-Khoury JM, Reineks EZ, Wang S (2011) Progress of liquid chromatography-mass spectrometry in measurement of vitamin D metabolites and analogues. Clin Biochem 44: 66–76
12. Couchman L, Benton CM, Moniz CF (2012) Variability in the analysis of 25-hydroxyvitamin D by liquid chromatography-tandem mass spectrometry: the devil is in the detail. Clin Chim Acta 413:1239–1243
13. Thibeault D, Caron N, Djiana R, Kremer R, Blank D (2012) Development and optimization of simplified LC-MS/MS quantification of 25-hydroxyvitamin D using protein precipitation combined with on-line solid phase extraction (SPE). J Chromatogr B 883–884:120–127
14. van den Ouweland JMW, Beijers AM, Demacker PNM, van Daal H (2010) Measurement of 25-OH-vitamin D in human serum using liquid chromatography tandem-mass spectrometry with comparison to radioimmunoassay and automated immunoassay. J Chromatogr B 878:1163–1168

15. Keevil BG (2013) Novel liquid chromatography tandem mass spectrometry (LC-MS/MS) method for measuring steroids. Best Prac Res Clin Endocrinol Metab 27:663–674
16. Fiers T, Casetta B, Bernaert B, Vandersypt E, Debock M, Kaufman JM (2012) Development of a highly sensitive method for the quantification of estrone and estradiol in serum by liquid chromatography tandem mass spectrometry without derivatization. J Chromatogr B 893–894:57–62
17. Kushnir MM, Rockwood AL, Bergquist J, Varshavsky M, Roberts WL, Yue B, Bunker AM, Meikle AW (2008) High-sensitivity tandem mass spectrometry assay for serum estrone and estradiol. Clin Chem 129:530–539
18. Keski-Rahkonen P, Huhtinen K, Poutanen M, Auriola S (2011) Fast and sensitive liquid chromatography-mass spectrometry assay for seven androgenic and progestagenic steroids in human serum. J Steroid Biochem Mol Biol 127:396–404
19. Shu I, Wang P (2013) Simultaneous serum nicotine, cotinine, and trans-3'-hydroxycotinine quantitation with minimal sample volume for tobacco exposure status of solid organ transplant patients. J Chromatogr B 928:139–145
20. Miller EI, Norris HRK, Rollins DE, Tiffany ST, Wilkins DG (2010) A novel validated procedure for the determination of nicotine, eight nicotine metabolites and two minor tobacco alkaloids in human plasma or urine by solid-phase extraction coupled with liquid chromatography-electrospray ionization-tandem mass spectrometry. J Chromatogr B 878:725–737
21. Baumann F, Regenthal R, Burgos-Guerrero IL, Hegerl U, Preiss R (2010) Determination of nicotine and cotinine in human serum by means of LC/MS. J Chromatogr B 878:107–111
22. Wishart DS, Tzur D, Knox C, Eisner R, Guo AC, Young N, Cheng D, Jewell K, Arndt D, Sawhney S, Fung C, Nikolai L, Lewis M, Coutouly MA, Forsythe I, Tang P, Shrivastava S, Jeroncic K, Stothard P, Amegbey G, Block D, Hau DD, Wagner J, Miniaci J, Clements M, Gebremedhin M, Guo N, Zhang Y, Duggan GE, MacInnis GD, Weljie AM, Dowlatabadi R, Bamforth F, Clive D, Greiner R, Li L, Marrie T, Sykes BD, Vogel HJ, Querengesser L (2007) HMDB: the human metabolome database. Nucleic Acids Res 35:D521–D526
23. Goldstein DS, Holmes C, Sharabi Y (2012) Cerebrospinal fluid biomarkers of central catecholamine deficiency in Parkinson's disease and other synucleinopathies. Brain 135 (6):1900–1913
24. Goldstein DS, Holmes C, Bentho O, Sato T, Moak J, Sharabi Y, Imrich R, Conant S, Eldadah BA (2008) Biomarkers to detect central dopamine deficiency and distinguish Parkinson disease from multiple system atrophy. Parkinsonism and related disorders 14: 600–607
25. Hou C, Jia F, Liu Y, Li L (2006) CSF serotonin, 5-hydroxyindolacetic acid and neuropeptide Y levels in severe major depressive disorder. Brain Res 1095:154–158
26. Vincent S, Bieck PR, Garland EM, Loghin C, Bymaster FP, Black BK, Gonzales C, Potter WZ, Robertson D (2004) Clinical assessment of norepinephrine transporter blockade through biochemical and pharmacological profiles. Circulation 109:3202–3207
27. Martin G, Mansion F, Houbart V, Paquet P, Rorive A, Chiap P, Crommen J, Servais AC, Fillet M (2011) Pre-study and in-study validation of a SPE-LC-MS-MS method for the determination of 5-S-cysteinyldopa, a melanoma biomarker, in human plasma. Talanta 84:280–286
28. Jian W, Edom R, Weng N, Zannikos P, Zhang Z, Wang H (2010) Validation and application of an LC-MS/MS method for quantitation of three fatty acid ethanolamides as biomarkers for fatty acid hydrolase inhibition in human plasma. J Chromatogr B 878:1687–1699
29. Penner N, Ramanathan R, Zgoda-Pols J, Chowdhury S (2010) Quantitative determination of hippuric and benzoic acids in urine by LC-MS/MS using surrogate standards. J Pharm Biomed Anal 52:534–543
30. Miliotis T, Lindberg C, Semb KF, van Geest M, Kjellstrom S (2013) Quantitative high-performance liquid chromatography-tandem mass spectrometry method for the analysis of free desmosines in plasma and urine. J Chromatogr A 1308:73–78

15. Kushnir ... (2013) Novel liquid chromatography tandem mass spectrometry (LC-MS/MS) method for measuring steroids. Best Pract Res Clin Endocrinol Metab [illegible]
16. Fiers T, Casetta B, Bernaert B, Vandersypt E, Debock M, Kaufman JM (2012) Development of a highly sensitive method for the quantification of estrone and estradiol in serum by liquid chromatography tandem mass spectrometry without derivatization. J Chromatogr B [illegible]
[illegible]

Application of Cell-Based Assays in Clinical Trials

Manjula P. Reddy and Amy Kate Sasser

Abstract Cell-based assays using primary cells and cell lines have been increasingly utilized to address scientific questions related to disease and therapeutic agents in clinical trials to: (1) gain greater understanding into the mechanism of action (MOA) of therapeutic agents; (2) identify patients that may respond to therapy; and (3) monitor pharmacodynamic biomarkers. Cell-based assays are also vital for the evaluation of target engagement or saturation, evaluation of safety markers, and monitoring mechanisms of resistance. However, implementation of cell-based assays in a clinical setting is challenging due to inherent variability (e.g., inter-subject, inter-assay) and stochastic factors that affect assay outcome. Hence, appropriate cell-based assay qualification and validation is required prior to testing clinical study samples. Assay qualification should include: (1) evaluation and optimization of relevant assay controls; (2) well-defined cell culture conditions; (3) stability determination of the analytes; and (4) inter-assay and intra-assay variance. This chapter provides an overview for some of the cell-based assays applicable for clinical studies. Examples of such assays include functional assays utilizing primary cells and cell lines, antibody-dependent cellular cytotoxicity (ADCC) and complement dependent cytotoxicity (CDC) assays, circulating tumor cell (CTC) assays, and receptor occupancy assays. Implementation and translational utility of these cell-based assays in clinical studies, including assay challenges and result interpretation, are discussed.

Keywords Cell-based assays · Antibody-dependent cellular cytotoxicity (ADCC) · Complement dependent cytotoxicity (CDC) · Circulating tumor cells (CTC) · ELISPOT

M.P. Reddy (✉) · A.K. Sasser
Oncology Translational Research, Janssen Research and Development,
Janssen Pharmaceutical Companies of Johnson and Johnson, LLC,
1400 McKean Road, Spring House, PA, USA
e-mail: Mreddy@its.jnj.com

A.K. Sasser
e-mail: KSasser1@ITS.JNJ.COM

R. Weiner and M. Kelley (eds.), *Translating Molecular Biomarkers into Clinical Assays*, AAPS Advances in the Pharmaceutical Sciences Series 21,
DOI 10.1007/978-3-319-40793-7_11

Functional Assays Using Primary Cells

ELISPOT Assay

The enzyme-linked immunosorbent spot (ELISPOT) assay is one of the most commonly used methods for T-cell monitoring in clinical studies. The utility of ELISPOT assays in detecting the frequency of antigen-specific T lymphocytes and immune responses in autoimmune, neoplastic, and infectious diseases has been reported in several studies [1–4]. This technology has been used for identification of T helper-1 (Th1)/T helper-2(Th2) effector class of the antigen-specific T-cell pool by detection of cytokines secreted by individual cells [5]. The frequencies of cytokine-producing cells cannot be obtained from other assays such as, enzyme-linked immunosorbent assay (ELISA) or reverse transcriptase polymerase chain reaction (RT-PCR) assays. Although in situ PCR does measure frequencies of cytokine positive cells, it does not perform well in the low-frequency range and measures mRNA (which frequently is posttranscriptionally controlled). Intracytoplasmic cytokine staining requires pharmacological disruption of the Golgi apparatus and usually additional signal-enhancing treatments to detect cytokines [6]. On the other hand, the cytokine ELISPOT assay measures the biologically relevant cytokine naturally released by individual cells over the entire duration of the assay. The advantage of ELISPOT assay is that it is highly quantitative and can measure broad range of cellular immune responses from T, B, and innate immune cells at a single cell level. These assays can be used to measure multiple secreted factors (e.g., cytokines, granzyme B) to assess cell function and to monitor treatment with immunotherapy agents. The adaptability of this assay in a clinical setting with evaluation of cryopreserved human lymphocytes without loss of functional responses and automated spot counting, contributed to the frequent use of this assay as a biomarker tool to predict or monitor clinical response following therapeutic immunomodulation [7, 8].

The ELISPOT method is based on the ELISA method, with some modifications. The ELISPOT assay method (Fig. 1) includes aseptically coating a capture antibody to the polyvinylidene difluoride (PVDF)-backed microwell plate followed by blocking the plate with serum proteins and then adding cells of interest, along with antigen or mitogen. A second antibody (biotinylated) is then added to detect the secreted cytokine captured on the PVDF membrane, which is then visualized using avidin-HRP and a precipitating substrate (e.g., AEC). The colored end product (spot) represents an individual cytokine-producing cell. The spots can be counted manually with a dissecting microscope, but commonly counted using automated spot readers. Results from clinical studies are typically represented as spot forming cells (SFC) per million cells. The ELISPOT method has emerged as a sensitive assay that can detect low frequency of antigen-specific T cells ($\leq$1 in 100,000) and is being increasingly used in clinical studies. However, as multiple factors can impact the assay readout, appropriate validation should be considered prior to clinical implementation, as previously described by Reddy et al. [9]. The most critical parameter to be standardized in the ELISPOT assay is cryopreservation and

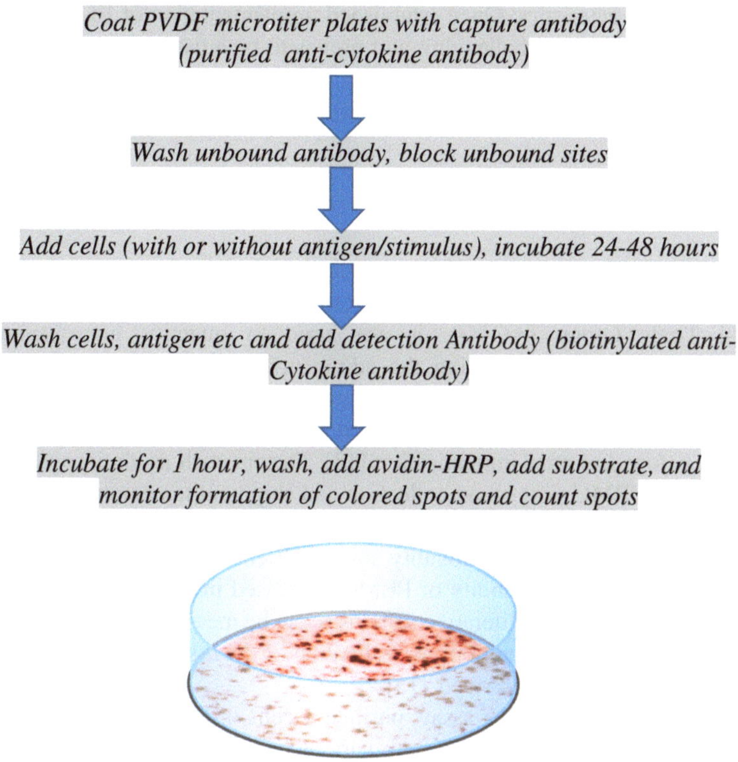

Fig. 1 ELISPOT assay procedure

thawing of peripheral blood mononuclear cells (PBMC) to ensure adequate recovery and viability [10–12]. Appropriate criteria for identifying a positive response, with inclusion of suitable assay controls, should be established along with relevant statistical tools for data analysis [9, 12]. Assay validation should include assessment of the stability of responses with sample shipment, cryopreservation along with intra-assay, inter-assay, and inter-operator variabilities. To minimize variability on assay readouts, it is recommended that samples from all time points from a clinical study subject be analyzed on the same day with appropriate plate controls, particularly for longitudinal monitoring of responses. Critical to establishing this assay for clinical studies is to ensure assay performance over time monitored by testing assay controls and other reagents. Cryopreserved control primary cell samples are also important for monitoring assay performance over time. Overall, adaptation of the ELISPOT assay in clinical studies requires development of a validated assay with preestablished criteria to identify a range of responses and a quality program that enables monitoring/tracking of assay performance characteristics. Once established, implementation of ELISPOT assays in clinical studies is less challenging and proves to be a useful tool for monitoring cellular responses.

Proliferation Assay

Antigen-specific T-cell proliferation is a major technique for assessing the functional capacity of CD4+ lymphocytes to respond to various stimuli. In the AIDS Clinical Trials Group (ACTG), it was used to quantify improvements in immunological function following the administration of a HIV-vaccine, and to detect the presence of immune responses against specific opportunistic pathogens. The proliferation assay has also been used in clinical studies to evaluate T-cell responses before and after immunization [13]. Factors such as, anticoagulant, time of isolation of PBMC post blood collection and shipment temperatures affect yield and viability of PBMC. Collection of blood in EDTA anticoagulant is reported to preserve antigen-specific functionality and proliferative responses [14]. In our laboratory, the yield and viability of PBMC decreased in blood samples processed after 48 h of collection. Proliferative responses can be measured as batch analysis in cryopreserved PBMC, but appropriate validation of the assay parameters should be performed as previously described [15].

Typically, in a proliferation assay, purified T cells or PBMC from clinical study subjects are cultured for 72–120 h with various concentrations of antigen, with or without stimulator cells, which are usually irradiated autologous or HLA-matched antigen-presenting cells. Tritiated thymidine is added during the last 8 h of culture to measure DNA synthesis (surrogate measure of proliferation). A stimulation index is then calculated by dividing the number of counts per minute (cpm) in the test well by the number of cpm in the wells with cells cultured without antigen (control). There are other nonradioactive commercially available proliferation assays (e.g., Alamar Blue or MTT) that are widely used. However, irrespective of the proliferation assay method, nonspecific immune function of the patients can influence the proliferation assay read outs, therefore inclusion of appropriate assay controls is vital to distinguish the response over background. A limitation of proliferation assays is that the stimulation index is not always reflective of the number of antigen-specific T cells present in vivo as high levels of proliferation by few cells or low levels of proliferation by many cells may result in a similar stimulation index. Other assay platforms such as, flow cytometry are also being used to measure proliferative responses. Flow cytometric assays that evaluate cell proliferation include: (i) measuring the distribution of cell membrane dyes into daughter cells to detect the number of cell divisions [16] and (ii) measuring cell cycle changes by assessing incorporation of bromodeoxyuridine into the DNA of the dividing cells [17]. However, flow cytometric assays are typically not suitable for large-scale testing of clinical samples and are best suited for real-time testing in smaller studies.

Functional Assays Using Cell Lines

Bioassays are typically used to estimate concentration or potency of a drug by measuring its biological responses in living systems. Bioassays commonly include a cell line that is responsive to a drug target leading to various biological responses such as, increase or decrease of proliferation, apoptosis, or secretion of a growth factor all of which can be measured using appropriate assay platforms (e.g., ELISA, flow cytometry). Cell-based bioassays are also used to determine if the biological effect of a therapeutic agent is neutralized by anti-drug antibody (ADA) response in patients [18]. Any bioassay used in clinical studies should be capable of fully characterizing the biological activity under investigation. Bioassays can be quantitative or qualitative. Development and validation of bioassays is highly challenging due to the variance of cell systems when challenged with biological agents.

Essential requirements for a bioassay are: (1) straight-forward design; (2) mimic biological functions in a reproducible manner; (3) demonstrate target specificity; and (4) tolerable to DMSO or other vehicles used for therapeutic reagent delivery. The bioassays should include appropriate positive and negative controls, and there should be minimal variability between reagent batches. Many bioassays also include a reference standard that normalizes the activity of the drug response being tested. A critical component of a bioassay is that the cell lines used in the assays is consistent throughout a clinical study. Ideally, biologic reagents should come from a single lot/source that has been well characterized, banked, and cultured following standard operating procedures. Validation of a bioassay should include evaluation of the common validation parameters such as accuracy, precision, stability, robustness, assay sensitivity, as well as, matrix effects. In addition, the assay should be reproducible with acceptable signal/noise readout and individual assay component variations must be minimized or controlled.

Assays to Evaluate Monoclonal Antibody Activity (ADCC and CDC)

Monoclonal antibodies are becoming more widely utilized across a variety of therapeutic applications. These therapeutics work through several mechanisms, including neutralization of the target antigen or inducing target cell death through a variety of mechanisms that require secondary participants in the forms of effector cells or complement. There are several in vitro assays that can be utilized to monitor the ability of the antibody to induce antibody-dependent cellular cytotoxicity (ADCC) or complement-dependent cytotoxicity (CDC) (Fig. 2).

ADCC assays utilize target cells that express the antigen of interest, as well as effector cells such as natural killer (NK) cells, monocytes, or leukocytes. The target cells are first incubated with the antibody to allow antibody: antigen binding and opsonization of the target cells. The effector cells are then added to induce target

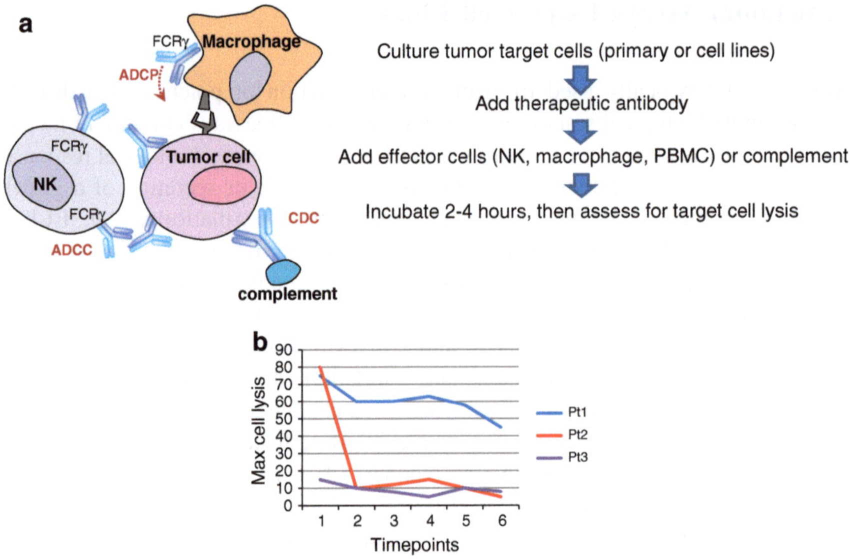

Fig. 2 Evaluating ADCC and CDC in clinical trial samples

cell killing. Target cells are typically labeled with ^{51}chromium, calcein-AM, or similar dye, to allow for measurement of target cell lysis. There are also enzymatic methods available, which do not require pre-labeling of target cells, but also do not distinguish between cell death of target cells or effector cells and therefore require additional controls. Some therapeutic antibodies induce not only effector cells killing of the target cell, but also effector: effector cell killing. In this scenario, a directly labeled target cell ADCC assay would allow effector and target cells to be characterized separately.

The various components of the ADCC assay can be modified depending on the research question. In oncology clinical studies, there are frequently questions regarding the sensitivity of the tumor cells to the therapeutic antibody. In this case, patient tumor samples may be collected and evaluated in an ADCC assay with a therapeutic antibody and a controlled effector cell population (either a cell line such as the NK-92 cell line, well characterized PBMC donor, or the patient's own effector cells). Measurements such as the effective concentration of a drug that gives half-maximal response (EC50) can be helpful in determining the sensitivity of the patient's tumor sample. This can be informative to monitor for initial response and potential development of resistance when a monoclonal antibody works primarily through ADCC. Patient samples can be collected at baseline and upon progression to determine whether the tumor cells have developed resistance to ADCC. Some challenges in this approach are that it can be difficult to collect large enough tumor cell numbers for complete evaluation, solid tumors require disaggregation for the assay, and liquid tumor biopsies are frequently not consistent in tumor cell percentages and may require presorting of tumor cells before in vitro assessment.

Therefore, these evaluations are best performed in a small controlled research environment or early phase 1–2 small-scale clinical studies and not in large global clinical trials. In another approach, the target cells can be held constant by utilizing a target cell line and the patient's effector cells can be monitored for their ability to carry out ADCC with a particular antibody of interest. These samples are more easily collected since it is typically a whole blood draw followed by PBMC isolation. These cells can be stored viably frozen until tested, and large cell numbers are typically readily available. In addition to these two clinical approaches, ADCC assays can be used in a high-throughput manner in drug development to screen multiple antibody candidates for their ability to induce targeted cell death.

The CDC assays follow a very similar format to ADCC (Fig. 2), but utilize purified complement proteins or serum containing these proteins in the place of effector cells. In a CDC assay, the antibody binds the antigen on the target cell followed by Clq complement factor binding with the CH_2 constant region of the antibody, which induces the formation of a membrane-attack complex (MAC) that lyses the target cell. CDC is a less complicated assay in the sense that the only cellular components are the target cells and the complement proteins do not interfere with measurement of tumor cell lysis. From a clinical trial perspective, it is also much easier to collect and store serum samples from patients compared to live cell collection, processing, and storage. One disadvantage is the long-term stability of complement proteins has not been established and these proteins are extremely labile. Similar to ADCC, the EC50 value can be reported for CDC assays, along with the maximum lysis for each sample analyzed. This data is frequently combined with direct measurement of specific complement proteins, to determine whether complement proteins are decreased in the subject after the therapeutic agent is delivered. And, similar to ADCC, these assays can be developed in a high-throughput manner for drug development [19].

In some cases, ADCC and CDC can be evaluated together in a single assay to monitor "target cell lysis" [20]. One example of this is in hematology oncology, where whole bone marrow aspirates are frequently used to assess tumor cell lysis in an ex vivo assay that assesses ADCC, CDC, and perhaps even antibody-dependent cellular phagocytosis (ADCP). Whole bone marrow aspirates can contain target tumor cells, NK effector cells (for ADCC), monocytes (for ADCP), and complement (for CDC), along with other effector cells, cross-linking antibodies, and several other potential mechanisms of target cell killing. While these combined assays are more reflective of the true in vivo biology, where multiple mechanisms may contribute to target cell death or the therapeutic agent's activity, they are also difficult to interpret when assessing the primary mechanism of action for the therapeutic agent. These assays are most useful when trying to understand target cell sensitivity and resistance to all potential mechanisms of action of a therapeutic agent [21].

Evaluation of Circulating Tumor Cells and Circulating Cell-Free (cf) Tumor DNA

Since 1869, it has been known that solid tumors shed cells into circulation and liquid tumors frequently have a circulating component, prompting the development of assays to detect and monitor circulating tumor cells (CTC). These assays operate under different principles depending on the platform, but the FDA approved CellSearch^{TM} platform captures tumor cells from whole blood via direct interaction of magnetic capture antibodies with antigens that are highly expressed on the tumor cell surface, such as EpCAM in solid epithelial tumors. After initial capture, tumor cells are further identified by negative selection and fluorescent labeling to remove all confounding white blood cells and non-tumor cells. The CellSearch platform is able to identify 1 CTC in 7.5 mL of whole blood, demonstrating greater sensitivity than standard flow cytometery methods. CTC are not present in normal healthy individuals allowing this assay to function as a diagnostic.

The CellSearch CTC assay was a significant independent predictor of progression-free survival (PFS) and overall survival (OS) in patients with metastatic breast, colorectal, or prostate cancer [22]. Although not yet prospectively validated, this assay may also be useful to predict patient response to therapy by monitoring baseline and posttreatment CTC counts. Additionally, it may identify patients with premalignancy or early disease states. This assay is easy to implement into clinical trials or clinical practice since it requires a single blood draw of 7.5 mL. One disadvantage is that there are currently validated assays available only for breast, colorectal, and prostate tumors, although the assay is also being evaluated for hematology approaches such as multiple myeloma (MM) and acute myeloid leukemia (AML). These assays require specialized and proprietary reagents and equipment that are available only at prespecified laboratories, and the test has only 96 h stability requiring immediate sample shipment to a testing laboratory. Current assays result in quantification of circulating tumor cells, but on-going clinical studies with exploratory assays are now investigating the ability of these platforms to go beyond enumeration and allow broader molecular characterization of captured tumor cells. Future assays will allow tumor cell gene expression profiling, sequencing, cytogenetic assessment, and deeper molecular characterization of circulating tumor cells which may aid further in selecting effective therapeutics and monitoring early response and/or resistance for oncology patients. In addition to capturing circulating tumor cells, tumor cell-free (cf) DNA is now also being captured and characterized as a potential oncology biomarker [23].

Receptor Occupancy Assays

Many therapeutic drugs, both small molecules and biologics, target specific receptors that drive downstream biological pathways. As part of the early drug development process, receptor occupancy is a key assay to evaluate the ability of a drug product to saturate or fully occupy a receptor and its resulting activity [24]. In most receptor occupancy assays, a ligand or drug product is labeled with a radioisotope or a fluorescent tag that can be quantified. This labeled drug product can be incubated with a tissue or cells that contain high levels of target, and compared to labeled natural ligand or a labeled noncompeting antibody that allows comparison of the drug product binding to total target expression. In the case of oncology clinical trials with therapeutic antibodies, tumor samples can be obtained at baseline and posttreatment to evaluate tumor cell expression of the target and the saturation or occupancy of the target at a particular dose and schedule of therapeutic antibody. These approaches typically use flow cytometry with labeled therapeutic antibody and a second noninterfering antibody that can measure the total target expression. The flow cytometric methods to measure receptor occupancy are described in more detail in Chapter "Key Mass Spectrometry Techniques used in Clinical Biomarker Research".

These assays are very useful in early drug development studies to determine the appropriate dose and schedule of the therapeutic drug. Most therapeutics that target a specific receptor aim to identify the dose and schedule that will shut down receptor activity, maximize on-target effects, and minimize off-target effects. Therefore, receptor occupancy is compared to pharmacokinetic (PK) measurements, clinical response, safety measurements, and side effect profiles to select final doses for further evaluation [25–27]. Receptor occupancy assays have advantages and disadvantages. They are extremely informative and helpful in understanding the biology of potential therapeutic agents, but require biopsies or whole blood draws at multiple time points and sufficient cells to perform these assays. In addition, specific reagents including labeled drug product and a noncompeting ligand or antibody are necessary.

Summary

- There is increasing interest in the use of cell-based in vitro assays to evaluate drug effects and mechanism of action, to monitor immune function, and to assess biomarkers of safety, prognosis, and clinical response.
- Cell-based assays offer the advantage of being closer to the human microenvironment and therapeutic targets. However, they are highly variable and require extensive validation and a standardized approach when utilized in clinical studies.

- There are several types of cell-based assays that are applicable for investigating a disease or response to a therapeutic agent. Selection of appropriate cell-based assays for clinical studies depends on the intended purpose of the assay, but any cell-based assay should be specific with appropriate assay controls to identify the effect of the test agent over assay background and other variables. Due to inherent variability in cell-based assays using primary cells, it is advisable to perform these assays in a central testing laboratory and in batch mode when possible.
- Newer platforms such as Cell Search, evaluation of cf DNA, combined immune-mediated cell toxicity assays, and multiparametric flow cytometry enable a more comprehensive assessment of several cellular parameters with higher sensitivity. As technologies improve and become more standardized, these methodologies can be more broadly implemented in clinical trials to inform biology and therapeutic drug development.

Acknowledgments The authors would like to thank Brett M. Hall, Senior Director Oncology Biomarkers, Janssen R&D for his review and suggestions. The authors would also like to thank the Diane Mohr, Director Biomarker Operations and Eric Dichter, Assistant General Counsel (all from Janssen R&D) for review and approval of the chapter.

References

1. Timothy MC, Amy CH, Paul JM, Lyerly KH, Morse AM (2001) Assays for monitoring cellular immune responses to active immunotherapy of cancer. Clin Cancer Res 7:1127–1135
2. Jansson AC, Ernerudh J, Kvarnstrom M, Ekerfelt C, Vrethem M (2000) Elispot assay detection of cytokine secretion in multiple sclerosis patients treated with interferon-beta1a or glatiramer acetate compared with untreated patients. Mult Scler 9(5):440–445
3. Helms T, Boehm BO, Asaad RJ, Trezza RP, Lehmann PV, Tary-Lehmann M (2000) Direct visualization of cytokine-producing recall antigen-specific CD4 memory T cells in healthy individuals and HIV patients. J Immunol 164:3723
4. Tary-Lehmann M, Hricik DE, Justice AC, Potter NS, Heeger PS (1998) Enzyme-linked immunosorbent assay spot detection of interferon-gamma and interleukin 5-producing cells as a predictive marker for renal allograft failure. Transplantation 66:219
5. Karulin AY, Hesse MD, Tary-Lehmann M, Lehmann PV (2000) Single-cytokine-producing CD4 memory cells predominate in type 1 and type 2 immunity. J Immunol 164:1862
6. Pala P, Hussell T, Openshaw PJ (2000) Flow cytometric measurement of intracellular cytokines. J Immunol Methods 243:107–124
7. Bennouna J, Hildesheim A, Chikamatsu K, Gooding W, Storkus WJ, and Whiteside TL (2002) Application of IL-5 ELISPOT assays to quantification of antigen-specific T helper responses. J Immunol Methods 261:145–156
8. Kreher CR, Dittrich MT, Guerkov R, Boehm BO, Tary-Lehmann M (2003) $CD4^+$ and $CD8^+$ cells in cryopreserved human PBMC maintain full functionality in cytokine ELISPOT assays. J Immunol Methods 278:79
9. Reddy M, Wong J, Pendley C, Prabhakar U (2008) IFN-γ ELISPOT assay validation. In: Validation of cell-based assays in the GLP setting, Chap. 10. John Wiley & Sons Ltd., pp 147–172

10. Ramachandran H, Laux J, Moldovan I, Caspell R, Lehmann PV, Subbramanian R (2012) Optimal thawing of cryopreserved peripheral mononuclear cells for use in high-throughput human immune monitoring studies. Cells 1:313–324
11. Kuerten S, Batoulis H, Recks MS, Karacsony E, Zhang W, Subbramanian R, Lehmann PV (2012) Resting of cryopreserved PBMC does not in general benefit the performance of antigen-specific T cell ELISPOT assays. Cells 1:313–324
12. Reddy M, Wong J, Davis H, Pendley C, Prabhakar U (2008) IL-5 ELISPOT assay validation. In: Validation of cell-based assays in the GLP setting, Chap. 11. John Wiley & Sons Ltd., pp 173–192
13. Disis ML, Grabstein KH, Sleath PR, Cheever MA (1999) Generation of immunity to the HER-2/neu oncogenic protein in patients with breast and ovarian cancer using a peptide-based vaccine. Clin Cancer Res 5:1289–1297
14. Kumar P, Satchidanandam V (2000) Ethyleneglycol-bis-(beta-aminoethylether)tetraacetate as a blood anticoagulant: preservation of antigen-presenting cell function and antigen-specific proliferative response of peripheral blood mononuclear cells from stored blood. Clin Diagn Lab Immunol 7(4):578–583
15. Reddy M, Prabhakar U (2008). Development optimization, and validation of cell based assays-2. Validation of cell-based assays in the GLP setting, Chap. 13. John Wiley & Sons Ltd., pp 25–36
16. Wallace PK, Tario JD, Fisher IL, Wallace SS, Ernstoff SS, Muirhead KA (2008) Tracking antigen-driven responses by flow cytometry: monitoring proliferation by dye dilution. Cytometry 73:1019–1034
17. Wilson GD (2000). Analysis of DNA-measurement of cell kinetics by the bromodeoxyuridine/anti-bromodeoxyuridine method. In: Ormerod MG (ed) Flow cytometry: a practical approach, 3rd edn. Oxford University Press, Oxford, pp 83–97
18. Rajadhyaksha M, Reddy M, Bald J, Fraunfelter A, Miller P, Kelly M, Prabhakar U (2008) Development of validated neutralization bioassays. In: Validation of cell-based assays in the GLP setting, Chap. 13. John Wiley & Sons Ltd., pp 209–237
19. Gerritsen AF, Bosch M, de Weers M, de Winkel JG, Parren PW (2010) High throughput screening for antibody induced complement-dependent cytotoxicity in early antibody discovery using homogeneous macroconfocal fluorescence imaging. J Immunol Methods 352(1–2):140–146
20. Wang SY, Weiner G (2008) Complement and cellular cytotoxicity in antibody therapy of cancer. Expert Opin Biol Ther 8(6):759–768
21. Van der Veer MS, De Weers M, Van Kessel B, Bakker JM, Wittebol S, Parren PW, Lokhorst HM, Mutis T (2011) Towards effective immunotherapy of myeloma: enhanced elimination of myeloma cells by combination of lenalidomide with the human CD38 monoclonal antibody daratumumab. Haematologica 96(2):284–290
22. Miller MC, Doyle GV, Terstappen LW (2010) Significance of circulating tumor cells detected by the CellSearch system in patients with metastatic breast colorectal and prostate cancer. J Oncol 617421–617428
23. Newman AM, Bratman SV, To J, Wynne JF, Eclov NC, Modlin LA, Liu CL, Neal JW, Wakelee HA, Merritt RE, Shrager JB, Loo BW Jr, Alizadeh AA, Diehn M (2014) An ultrasensitive method for quantitating circulating tumor DNA with broad patient coverage. Nat Med 20(5):548–554
24. Grimwood S, Hartig PR (2009) Target site occupancy: emerging generalizations from clinical and preclinical studies. Pharmacol Ther 122(3):281–301
25. Waibler Z, Sender LY, Kamp C, Müller-Berghaus J, Liedert B, Schneider CK, Löwer J, Kalinke U (2008) Toward experimental assessment of receptor occupancy: TGN1412 revisited. J Allergy Clin Immunol 122(5):890–892

26. Ma A, Dun H, Song L, Hu Y, Zeng L, Bai J, Zhang G, Kinugasa F, Miyao Y, Sakuma S, Okimura K, Kasai N, Daloze P, Chen H (2014) Pharmacokinetics and pharmacodynamics of ASKP1240, a fully human anti-CD40 antibody, in normal and renal transplanted cynomolgus monkeys. Transplantation 97(4):397–404
27. Goldwater R1, Keirns J, Blahunka P, First R, Sawamoto T, Zhang W, Kowalski D, Kaibara A, Holman J (2013) A phase 1 randomized ascending single-dose study of antagonist anti-human CD40 ASKP1240 in healthy subjects. Am J Transplant 13(4):1040–1046

New Technologies for Cellular Analysis

Peter J. O'Brien, Tim Wyant and Virginia Litwin

Abstract Cytometric technologies have been indispensable for understanding biological and pathological processes, and are increasingly used to provide critical information on safety and efficacy in drug development. Highly sophisticated multiparametric cytometry methods are now available to measure treatment-induced changes in the phenotypes and functions of individual cells in heterogeneous populations. Numerous phenotypic and functional cytometry assays have been validated for pharmacodynamic studies in clinical drug trials, and that number is likely to expand as new analytical technologies become available. This chapter will discuss three new cytometric technologies that will likely impact clinical drug development in the near future: Imaging cytometry on a chip; Imaging flow cytometry; and Mass cytometry. Each of these platforms is well-suited to specific aspects of cellular analysis, and combines new technologies with tried and true cytometry methods.

P.J. O'Brien
Department of Pharmacokinetics, Dynamics and Metabolism, Pfizer, 10646 Science Center Drive, San Diego, CA 92121, USA
e-mail: Peter.O'Brien2@pfizer.com

T. Wyant
Department of Translational Medicine, Curis Inc., 4 Maguire Rd, Lexington, MA 02421, USA
e-mail: Twyant@curis.com

V. Litwin (✉)
Department of Hematology/Flow Cytometry, Covance Inc., 8211 SciCore Drive, Indianapolis, IN 46214, USA
e-mail: virginia.litwin@covance.com

R. Weiner and M. Kelley (eds.), *Translating Molecular Biomarkers into Clinical Assays*, AAPS Advances in the Pharmaceutical Sciences Series 21,
DOI 10.1007/978-3-319-40793-7_12

Key Terms

Depth of field (DOF) The optical distance across which objects are acceptably sharp and in focus varies depending upon microscope hardware and the method of image acquisition. Traditional IFM uses lenses with relatively narrow DOF, whereas in confocal microscopy, samples can be "optically sectioned" and reassembled to represent three dimensional cellular structures in sharply focused two-dimensional images. The latter process is relatively slow and suffers from photobleaching and other untoward effects of sample reanalysis that are required for monitoring intracellular changes over time

Hydrodynamic Focusing In flow cytometry, individual cells in suspension are analyzed. Most flow cytometers rely on a process call hydrodynamic focusing to direct single cells for interrogation to the laser light source. Briefly, the cell suspension is contained in a stream of fluid centered within an outer stream. The two fluids differ enough in their velocity and form a two-layer stable flow. Within the laminar flow, the cells orient with their long axis parallel to the flow

Fluorescence spectral overlap Fluorochromes are excited at one wavelength of light and emit energy at another. The histogram display of the emission spectra from various fluorochromes shows a major peak indicating the wave length where most of the signal will resolve and a shoulder or tail where a smaller portion of the signal can be detected. The optics of a flow cytometer are setup such that the major signal from each fluorochrome is detected in a specific channel. A situation where a small portion of the signal from one fluorochrome overlaps with the detection channel of a second fluorochrome is referred to as fluorescence spectral overlap

Chip Cytometry and Cell Imaging

Laser scanning cytometry (LSC) is an established method for quantifying the fluorescence of immobilized cells, and has been used to characterize patient treatment responses in clinical trials [1]. A variety of LSC platforms have been developed that interrogate cells mounted on slides or cartridges containing the sample. For example, the Imagn2000 instrument from Biometric Imaging has been used to enumerate cluster of differentiation (CD) CD4 and CD8 lymphocytes during treatment of human immunodeficiency virus (HIV) infected patients [2]. This instrument uses a simple 2-color-Photomultiplier Tube (PMT) system and a slide-based cartridge to interrogate cells while minimizing the risk from aerosols. SurroMed markets the SurroScan, a 4-color LSC instrument that uses cartridges with 32 wells for absolute cell counting in small volumes of whole or diluted blood. The CompuCyte iCyte® LSC uses advanced optics and CCD cameras for collection of four fluorescence channels and optical images of cells. This approach allows the examination of smaller quantities of blood while sparing reagents, and may reduce artifactual cell activation in unfixed samples, two useful characteristics for pharmacodynamics studies [3, 4].

Traditional flow cytometry (FC) functional assays can be adapted for LSC, although the total number of cells acquired in LSC is considerably smaller than can be achieved using more traditional flow-based methods. Also, as most LSC instruments are limited to four colors, the detection of rare subpopulations can be difficult, if not impossible using LSC. In further contrast to flow cytometers, which use hydrodynamic focusing to align cells, microfluidic flow cytometers like the Fishman-R and instruments by Zellkraftwerk (also known as "chip cytometers") pass cells through micro-fabricated channels etched onto chips, where they are illuminated by lasers for the measurement of cellular fluorescence. Such miniaturization reduces required sample volumes and reagent costs, and allows the collection of both cytometric data and cell images [5].

Complications associated with low cell numbers and inadequate numbers of fluorescence channels can be further aggravated by chip-to-chip variations in microfluidics channels and optical properties. The Zellkraftwerk instrument [6] is one of the newest iterations of chip cytometers, and seeks to address these issues via a combination of LSC technology, advanced microfluidics, and methods for staining and restaining of cells in a sample (Fig. 1) [7]. Similar to traditional LSC, this instrument captures and scans cells on a fixed slide (Fig. 2). Re-staining of cells is accomplished by serially measuring stained cell fluorescence, then photo bleaching the initial sample to allow re-staining with noncompetitive antibodies for

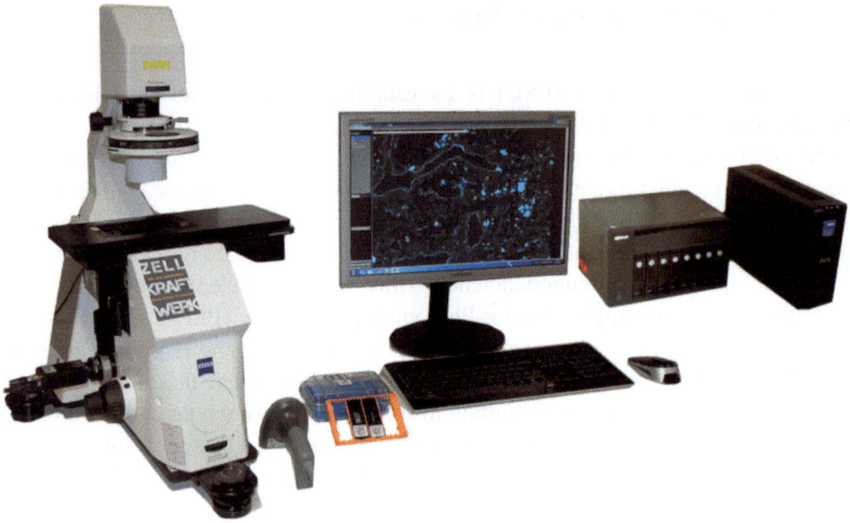

Fig. 1 Zellscanner ONE. Reproduced with kind permission from Zellkraftwerk

measurement of additional antigens. Thus, despite the limited number of fluorochromes that can be measured using this instrument, the ability to re-assay samples expands the effective number of parameters queried. It is important to consider the photo stability of a given fluorophore when designing re-staining protocols, since many of the newer dyes are designed to be resistant to photo bleaching.

Some characteristics of these platforms are quite useful for clinical trials support, but other factors serve to limit their utility. For example, the narrow field of view and low sample volumes used in these methods means that these instruments generally have a reduced capacity for rare event detection, which can be required for determining minimal residual disease status in hematological malignancies. Furthermore, in the absence of additional manufacturing quality controls, inter-chip variability may limit the reproducibility of chip cytometry in longitudinal pharmacodynamic studies. Platform availability at contract labs can also present challenges for clinical trials support. Individual contract labs may be sufficient for smaller clinical trials, but as a drug progresses through development, the need for additional instruments and greater regulatory oversight become paramount, and may require the participation of larger contract labs.

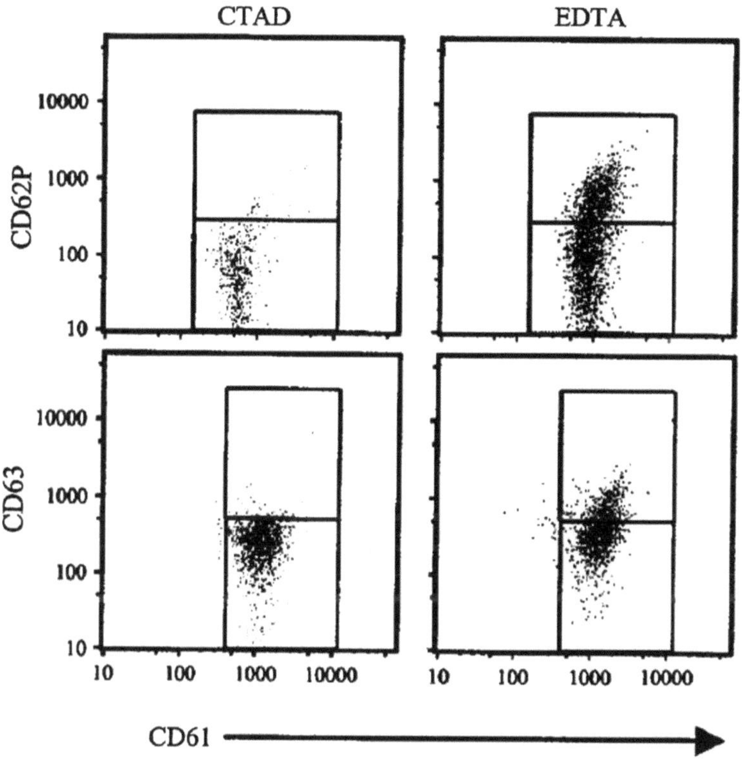

Fig. 2 The use of LSC to measure resting and activated platelets. Platelet rich plasma was prepared from either EDTA or CTAD blood and CD61+ platelets were examined for the expression of the platelet activation markers CD63 and C62P. Reproduced from Wyant et al. with permission

Imaging Flow Cytometry

Traditional FC and automated immuno-fluorescence microscopy (IFM) have proven useful for phenotypic screening and multiparametric cell profiling in early phase drug development [8]. FC and IFM can also improve our understanding of exposure-response relationships in animal studies and clinical trials, but their respective methodological advantages are to a large extent mutually exclusive. FC, for example, provides reliable information regarding the bulk fluorescence of specifically gated cell populations, but has relatively low single-cell resolution, and provides limited spatial information about biomarker micro-anatomical distribution, relative abundance, or normalized activation status, limiting our understanding of complex biological signaling. IFM provides high resolution fluorescence data, and can be used to monitor cell morphology and localize fluorescence signals in specific cells over time. Though amenable to automation, IFM has much lower throughput and provides less statistical power than FC, offers a limited capacity for analyzing suspension cells, suffers from susceptibility to photo bleaching, and is subject to observer bias through the selection of visual fields. IFM systems also typically offer a limited number of simultaneous excitation sources and imaging modes, limiting the size of multiplex panels.

Imaging flow cytometry (IFC) is a hybrid method that combines the statistical power of multiparametric FC with the spatial and morphological discrimination of fluorescence microscopy, enabling the simultaneous capture of multi-mode imagery (i.e., bright field, dark field, and fluorescence images) [9]. Excellent reviews of IFC have recently been published that provide a good introduction to this technology and specific applications of relevance to drug development [10–14]. Other publications provide useful comparisons of flow imaging with static methods like the CellTracks® system [Johnson and Johnson] [15], highlighting the key technological advances that have driven the broad acceptance of IFC as a tool for both basic and applied research. Over 350 publications describe IFC studies, and the number of novel applications of IFC in basic research is steadily increasing. The utility of IFC is further exemplified in the widespread adoption of imaging cytometers made by Amnis/EMD Millipore, who ten years ago marketed their first of several generations of multispectral imaging cytometers (Fig. 3) [16].

In IFC cells are loaded in suspension into a hydrodynamically focused fluid stream that passes through a flow cell for illumination and detection of specific spectral and morphological characteristics. Here, key differences between FC and IFC begin to emerge. First, in IFC sample loading speed and volume are carefully

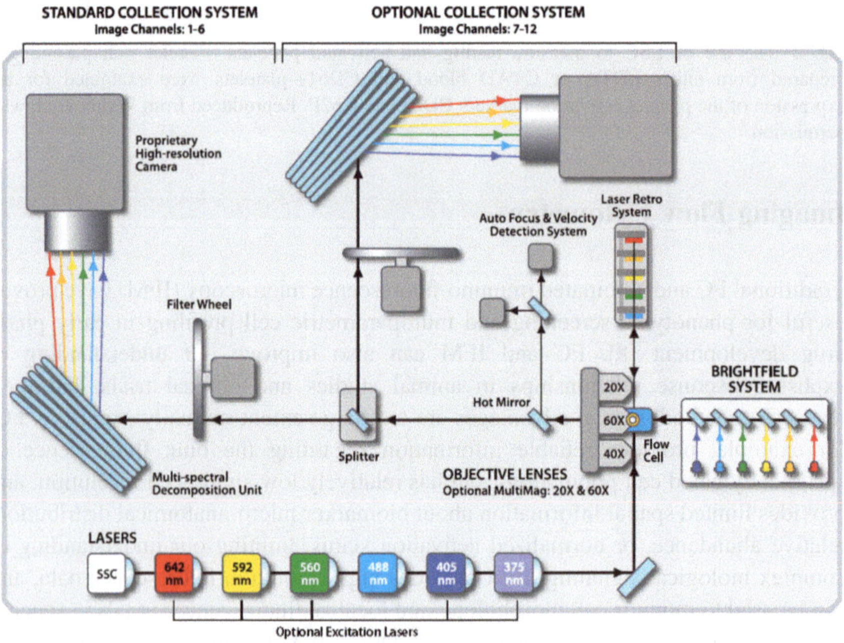

Fig. 3 Schematic of Amnis. Reproduced with kind permission from EMD Millipore

controlled to allow synchronized, time delayed integration (TDI) of optical signals generated after illumination with a bright field light source and at least one laser. In this approach, cells are tracked and images are captured by panning across the flow cell using a high numerical aperture objective. Transmitted and scattered light and cellular fluorescence are captured and spectrally deconvoluted on multichannel CCD cameras. Alignment of these data allows the simultaneous capture of fluorescence data and sharp images of each cell, allowing unbiased quantification of spatial information about target molecules in snapshots of individual cells (Fig. 4).

IFC allows refinements in gating strategies to include cellular aspect ratio, cell diameters, and cell volumes in addition to the traditional light scattering and fluorescence-based gating used in FC. In IFC, individual events are also captured as static visual images that can be integrated with fluorescence data at relatively high resolution, providing an essentially infinite number of options for cell classification, and allowing definitive visual confirmation of event gating. Newer Amnis® IFC instruments can be modified to provide additional functions like extended depth of field imaging for detecting fluorescent puncti over a wider focal range, a useful feature for studies of DNA ploidy and cell-cell interactions [13, 16–18]. This combination of features provides superior discrimination of subtly different populations in heterogeneous cell mixtures.

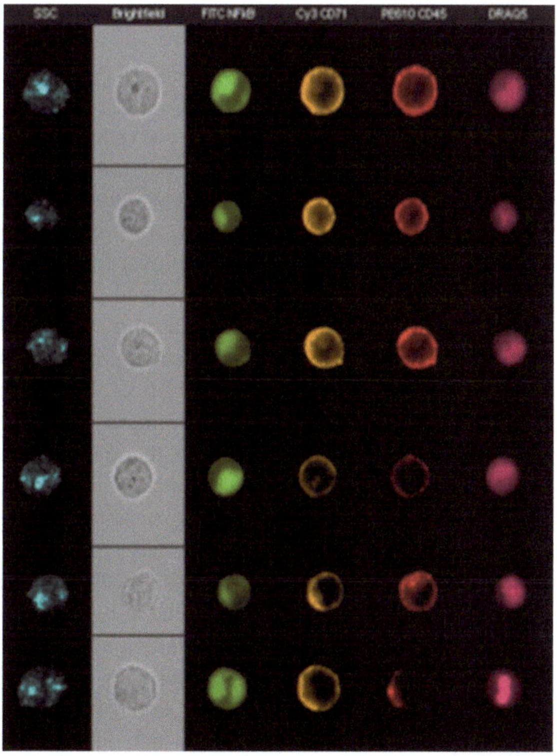

Fig. 4 Amnis cellular images. Reproduced with kind permission from EMD Millipore

IFC reagent selection, sample preparation and data acquisition are very similar to that of traditional FC, meaning that established FC methods are often readily adaptable to IFC. Numerous functional FC assays have been modified for IFC, including the quantification and correlation of shape change and other morphological characteristics with the cell cycle and other cellular events [11, 19]; phagocytosis assays [20–22]; analysis of DNA damage and repair and other events associated with cell death and autophagy [23]; cell-cell and cell-particle interactions and the exchange of cytoplasmic contents [24]; co-localization of intracellular and cell surface epitopes [25]; monitoring of protein interactions and trafficking to organelles [25]; and spot counting for ploidy determination and other applications of in situ hybridization [26]. Though suitable for discovery research and many academic purposes, most published IFC applications are only preliminarily validated, and can be poorly suited to drug development clinical trials. For example, IFC data acquisition is slower than traditional cytometry, and under the most common configurations IFC data files can be hundreds of times larger than files from a traditional flow cytometer. The learning curve for IFC data analyses is significant for those without experience in advanced microscopy, and the high cost of purchasing multiple IFC instruments puts backup instruments out of reach of most labs. This is a critical hurdle for drug development, since backup instruments can be required for regulatory compliance in support of clinical trials.

CyTOF Mass Cytometry

In the past decade, technological advances in fluorescent probes, cytometry instrumentation, and data analysis software have enabled "high dimensional flow cytometry" where 20 parameters (18 fluorescent-labeled probes and two light scatter properties) of data can be collected from an individual cell. High-dimensional flow cytometry allows for a far more in depth cellular characterization and dissection of more refined cellular subsets. Indeed, the ability to measure up to 20 parameters of data on a single cell has been critical in enabling advances in research focusing on hematopoiesis, complex immune responses, and intracellular regulatory signaling networks. Using the existing instrumentation and fluorescent probes, the number of parameters is unlikely to increase beyond twenty. This upper limit is largely due to the overlap in the emission spectra of the fluorescent probes resulting in the detection of one fluorophore in multiple detector channels and the challenges in compensating for this overlap. An innovative, relatively new technology, mass cytometry, or CyTOF® [Fluidigm], can extend the capability of highly multiparametric analysis well beyond 20; already studies using more than thirty-parameter analysis have been published [27].

CyTOF® mass cytometry, essentially a hybrid between flow cytometry (Cy) and time-of-flight (TOF) mass spectrometry, is based on the concept that isotopically pure heavy metal reporter elements could be conjugated to cellular probes (most commonly monoclonal antibodies) which could then be quantified in an inductively coupled plasma mass spectrometry (ICP-MS) detection system [27]. CyTOF® provides at least three orders of magnitude of resolution between adjacent detection channels, thus the use of heavy metal probes rather than fluorescent-labeled probes practically eliminates the need for compensation, removing the parameter restrictions and other technical challenges associated with fluorescence spectral overlap compensation.

CyTOF® mass cytometry has a workflow somewhat similar to that of high-dimensional flow cytometry in that the labeled heterogeneous populations of cells are individually analyzed. Unlike flow cytometry where single cells are interrogated by a laser light source, in mass cytometry cells are nebulized into single-cell droplets and introduced into the plasma, where they are completely vaporized into component elemental ions (Fig. 5) [28]. The cloud of atomic ions for each single cell is extracted into the ion optics and time-of-flight regions of the mass cytometer where the ions are separated by mass. To resolve the probe ions from the abundant cellular and antibody ions, the mass cytometer is configured as a quadrupole–time-of-flight (qTOF) instrument. The quadrupole acts as a filter allowing only the heavier ions (probe) to be quantitated by TOF mass analysis. The masses corresponding to the metal-tagged probes are counted in discrete time-separated detector channels reminiscent of fluorescence emission detection in

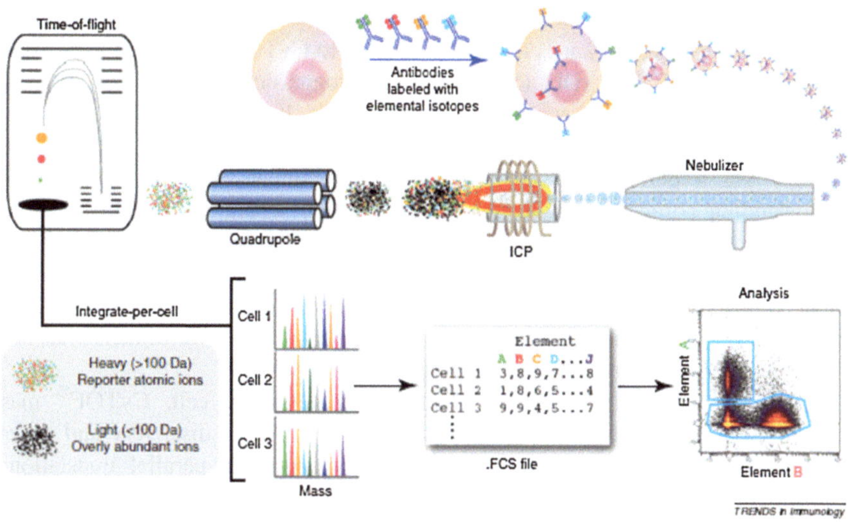

Fig. 5 Schematic of ICP-MS-based analysis of cellular markers. An affinity product (e.g., antibody) tagged with a specific element binds to the cellular epitope. The cell is introduced into the ICP by droplet nebulization. Each cell is atomized, ionized, overly abundant ions removed, and the elemental composition of remaining heavy elements (reporters) is determined. Signals corresponding to each elemental tag are then correlated with the presence of the respective marker and analyzed using conventional cytometry platforms. Reprinted from Bendall et al. [28], Copyright (2012), with permission from Elsevier

the appropriate detector PMT. Much as fluorescence emission is proportional to the level of antibody binding, the intensity of the heavy metal signal detected in each channel is directly proportional to the number of specific probe-derived ions striking the detector and thus the number of antibodies originally bound per cell.

As with any new and highly novel technology, the challenges with CyTOF® mass cytometry are only beginning to be identified. To date, there are a limited number of formal comparisons of mass cytometry and polychromatic flow cytometry [30] and issues regarding sensitivity for some surface antigens and cell loss rates during acquisition are beginning to be discussed [31, 32]. Minor challenges regarding reagent availability are likely to decrease as the technology is more widely utilized. The primary challenge is that currently there are no quality assurance and normalization protocols [33]. The technology lacks high-throughput capabilities and changes in instrument performance are evident after a few hours of acquisition. Between run fluctuations have been reported. Instrument standardization and monitoring are essential in order for this technology to be of value outside of the basic sciences research arena and drug screening applications. A normalization algorithm based on prominent features or "landmarks" in raw flow cytometry data was recently used to correct for instrument variability [34]. Another challenge is the organization and analysis of the high-dimensional data. Although the files are saved in the .fcs file format to allow gating in any flow cytometry data analysis package, traditional methods of sequential, Boolean gating would not allow maximum utilization of high-dimensional data sets. Fortunately, several interesting comparative multivariate analysis packages have been applied to CyTOF® data such as SPADE, PAC, and viSNE. SPADE (Spanning-tree Progression Analysis of Density-normalized Events) [35], was the first to appear. It is a clustering algorithm, which allows identification of low-density clusters and displays the relatedness of clusters via a dendrogram [35]. Principle components analysis (PCA), is a long-standing computational technique, [36] which separates a group of events according to their measured attributes and has recently been applied to CyTOF® data. PCA allows for the clustering of cells that are phenotypically distinct from other cells. viSNE is a recently described algorithm for high-dimensional data analysis [37] in which individual events are displayed on a two-dimensional map which preserves the multi-dimensional separation [38, 39].

With the large number of parallel measurements per cell, CyTOF® mass cytometry is potentially a very powerful new tool for drug discovery and development. It could allow for the identification of multiple parallel translational pathway responses to agonist/antagonist intervention. In the past, high-dimensional flow cytometry was sometimes referred to as proteomics at the single cell level; with the potential to measure 100 parameters per cell CyTOF® mass cytometry would more closely achieve that goal. It is easy to imagine that the technology might be used for hypothesis generating experiments much in the way as gene chips. For example, samples collected before and after therapeutic intervention could be stained with a variety of CD markers for analysis of changes in the clustering patterns.

Conclusions

FC is considered to be the optimal technology for the analysis of large numbers of heterogeneous cellular populations. In the drug development process, FC has been applied to drug screening and lead compound characterization, preclinical biomarker and pharmacodynamic studies, and to patient stratification drug response outcomes during clinical trials [40]. The value of established cytometry platforms in many of these applications is increasingly obvious and, once they are properly validated, some of the technologies described here are likely to help fill existing gaps in experimental methodologies and instrumentation. The challenges before us may seem daunting, but as was the case for FC assays these past two decades, it is likely that the emergence of additional novel technologies will aid the evolution of established methods onto new cytometry platforms. The improved ability to characterize immune cell phenotypes and functional responses to therapy afforded by these emerging technologies is likely to spur additional advances in cancer immunotherapy, autoimmunity research, and the diagnosis and treatment of chronic viral diseases. As cytometers continue to shrink [41] and cytometry expands into the realm of molecular diagnostics [27], it is increasingly likely that cytometry will play a major role in the optimization of personalized drug therapies and health care delivery in lesser developed nations.

Summary Box

- Cytometric technologies are indispensable for understanding biological and pathological processes, and are increasingly used to provide information on safety and efficacy in drug development.
- Highly sophisticated multiparametric cytometry methods are now available to measure treatment-induced changes in the phenotypes and functions of individual cells in heterogeneous populations.
- Laser scanning cytometry (LSC) is an established method for quantifying the fluorescence of immobilized cells, and has been used to characterize patient treatment responses in clinical trials.
- Microfluidic flow cytometers like the Fishman-R and instruments by Zellkraftwerk (also known as "chip cytometers") pass cells through micro-fabricated channels etched onto chips, where they are illuminated by lasers for the measurement of cellular fluorescence. Such miniaturization reduces required sample volumes and reagent costs, allows the collection of both cytometric data and cell images, and may reduce artifactual cell activation in unfixed samples.
- Imaging flow cytometry (IFC) is a hybrid method that combines the statistical power of multiparametric FC with the spatial and morphological discrimination of fluorescence microscopy.

- IFC allows refinements in gating strategies to include cellular aspect ratio, cell diameters, and cell volumes in addition to the traditional light scattering and fluorescence-based gating used in FC.
- CyTOF® mass cytometry, essentially a hybrid between flow cytometry (Cy) and time-of-flight (TOF) mass spectrometry, is based on the concept that isotopically pure heavy metal reporter elements could be conjugated to cellular probes (most commonly monoclonal antibodies) which could then be quantified in an inductively coupled plasma mass spectrometry (ICP-MS) detection system.
- CyTOF® provides at least three orders of magnitude of resolution between adjacent detection channels, thus the use of heavy metal probes rather than fluorescent-labeled probes practically eliminates the need for compensation, removing the parameter restrictions and other technical challenges associated with fluorescence spectral overlap compensation.

References

1. Jonasch E, McCutcheon IE, Waguespack SG et al (2011) Pilot trial of sunitinib therapy in patients with von Hippel-Lindau disease. Ann Oncol Official J Eur Soc Med Oncol/ESMO 22(12):2661–2666
2. Shapiro HM (2004) Scanning laser cytometry. Current protocols in cytometry/editorial board, Robinson JP, managing editor et al Chapter 2, Unit 2 10
3. Kantor AB, Alters SE, Cheal K, Dietz LJ (2004) Immune systems biology: immunoprofiling of cells and molecules. Biotechniques 36(3):520–524
4. Wyant TL, Smith PC, Brown B, Kantor AB (2001) Whole blood microvolume laser scanning cytometry for monitoring resting and activated platelets. Platelets 12(5):309–318
5. Godin J, Chen CH, Cho SH, Qiao W, Tsai F, Lo YH (2008) Microfluidics and photonics for Bio-System-on-a-Chip: a review of advancements in technology towards a microfluidic flow cytometry chip. J Biophotonics 1(5):355–376
6. www.zellkraftwerk.com; www.zellkraftwerk.com
7. Hennig C, Adams N, Hansen GA (2009) Versatile platform for comprehensive chip-based explorative cytometry. Cytometry Part A J Int Soc Anal Cytol 75(4):362–370
8. Feng Y, Mitchison TJ, Bender A, Young DW, Tallarico JA (2009) Multi-parameter phenotypic profiling: using cellular effects to characterize small-molecule compounds. Nat Rev Drug Discovery 8(7):567–578
9. Rimon N, Schuldiner M (2011) Getting the whole picture: combining throughput with content in microscopy. J Cell Sci 124(Pt 22):3743–3751
10. Basiji DA, Ortyn WE, Liang L, Venkatachalam V, Morrissey P (2007) Cellular image analysis and imaging by flow cytometry. Clinics Lab Med 27(3):653–670, viii
11. McGrath KE, Bushnell TP, Palis J (2008) Multispectral imaging of hematopoietic cells: where flow meets morphology. J Immunol Methods 336(2):91–97
12. Barteneva NS, Fasler-Kan E, Vorobjev IA (2012) Imaging flow cytometry: coping with heterogeneity in biological systems. The J Histochem Cytochem Official J Histochem Soc 60 (10):723–733
13. Payes C, Rodriguez JA, Friend S, Helguera G (2012) Cell Interaction Analysis by Imaging Flow Cytometry. In: Gowder S (ed) Cell interaction (InTech, Rijeka, Croatia, 2012), p 303–322

14. Elliott GS (2009) Moving pictures: imaging flow cytometry for drug development. Comb Chem High Throughput Screening 12(9):849–859
15. Scholtens TM, Schreuder F, Ligthart ST et al (2011) CellTracks TDI: an image cytometer for cell characterization. Cytometry. Part A J Int Soc Anal Cytol 79(3):203–213
16. Zuba-Surma EK, Ratajczak MZ (2011) Analytical capabilities of the ImageStream cytometer. Methods Cell Biol 102:207–230
17. George TC, Basiji DA, Hall BE et al (2004) Distinguishing modes of cell death using the ImageStream multispectral imaging flow cytometer. Cytometry Part A J Int Soc Anal Cytol 59(2):237–245
18. Ortyn WE, Perry DJ, Venkatachalam V et al (2007) Extended depth of field imaging for high speed cell analysis. Cytometry Part A J Int Soc Anal Cytol 71(4):215–231
19. Filby A, Perucha E, Summers H et al (2011) An imaging flow cytometric method for measuring cell division history and molecular symmetry during mitosis. Cytometry Part A J Int Soc Anal Cytol 79(7):496–506
20. Ackerman ME, Moldt B, Wyatt RT et al (2011) A robust, high-throughput assay to determine the phagocytic activity of clinical antibody samples. J Immunol Methods 366(1–2):8–19
21. Ahmed F, Friend S, George TC, Barteneva N, Lieberman J (2009) Numbers matter: quantitative and dynamic analysis of the formation of an immunological synapse using imaging flow cytometry. J Immunol Methods 347(1–2):79–86
22. Ploppa A, George TC, Unertl KE, Nohe B, Durieux ME (2011) ImageStream cytometry extends the analysis of phagocytosis and oxidative burst. Scand J Clin Lab Invest 71(5): 362–369
23. de la Calle C, Joubert PE, Law HK, Hasan M, Albert ML (2011) Simultaneous assessment of autophagy and apoptosis using multispectral imaging cytometry. Autophagy 7(9):1045–1051
24. De Cuyper IM, Meinders M, van de Vijver E et al (2013) A novel flow cytometry-based platelet aggregation assay. Blood 121(10):e70–e80
25. Gilcrease MZ, Zhou X, Lu X, Woodward WA, Hall BE, Morrissey PJ (2009) Alpha6beta4 integrin crosslinking induces EGFR clustering and promotes EGF-mediated Rho activation in breast cancer. J Exp Clin Cancer Res CR 28:67
26. George TC, Fanning SL, Fitzgerald-Bocarsly P et al (2006) Quantitative measurement of nuclear translocation events using similarity analysis of multispectral cellular images obtained in flow. J Immunol Methods 311(1–2):117–129
27. Minderman H, Humphrey K, Arcadi JK et al (2012) Image cytometry-based detection of aneuploidy by fluorescence in situ hybridization in suspension. Cytometry Part A J Int Soc Anal Cytol 81(9):776–784
28. Bandura DR, Baranov VI, Ornatsky OI, Antonov A, Kinach R, Lou X, Pavlov S, Vorobiev S, Dick JE, Tanner SD (2009) Mass cytometry: technique for real time single cell multitarget immunoassay based on inductively coupled plasma time-of-flight mass spectrometry. Anal Chem 81(16):6813–6822
29. Bendall SC, Nolan GP, Roederer M, Chattopadhyay PK (2012) A deep profiler's guide to cytometry. Trends Immunol 33(7):323–332
30. Wang L, Abbasi F, Ornatsky O et al (2012) Human CD4+ lymphocytes for antigen quantification: characterization using conventional flow cytometry and mass cytometry. Cytometry Part A J Int Soc Anal Cytol 81(7):567–575
31. http://cytoforum.stanford.edu; http://www.webcitation.org/6NMOMZ3WM
32. cytometry@lists.purdue.edu; http://www.webcitation.org/6NMOY4SMU
33. Finck R, Simonds EF, Jager A et al (2013) Normalization of mass cytometry data with bead standards. Cytometry Part A J Int Soc Anal Cytol 83(5):483–494
34. Hahne F, Khodabakhshi AH, Bashashati A et al (2010) Per-channel basis normalization methods for flow cytometry data. Cytometry Part A J Int Soc Anal Cytol 77(2):121–131
35. https://dvs.cytobank.org; http://www.webcitation.org/6NMObooLR
36. Qiu P, Simonds EF, Bendall SC et al (2011) Extracting a cellular hierarchy from high-dimensional cytometry data with SPADE. Nat Biotechnol 29(10):886–891

37. Newell EW, Sigal N, Bendall SC, Nolan GP, Davis MM (2012) Cytometry by time-of-flight shows combinatorial cytokine expression and virus-specific cell niches within a continuum of CD8+ T cell phenotypes. Immunity 36(1):142–152
38. el·Amir AD, Davis KL, Tadmor MD et al (2013) viSNE enables visualization of high dimensional single-cell data and reveals phenotypic heterogeneity of leukemia. Nat Biotechnol 31(6):545–552
39. http://www.c2b2.columbia.edu/danapeerlab/html/cyt.html; http://www.webcitation.org/6NMMnoZ6k
40. O'Hara DM, Xu Y, Liang Z, Reddy MP, Wu DY, Litwin V (2011) Recommendations for the validation of flow cytometric testing during drug development: II assays. J Immunol Methods 363(2):120–134
41. Wouters FS, Wessels JT (2012) Cell phones go cellular–current scale-down lab-in-your-pocket applications. Cytometry Part A J Int Soc Anal Cytol 81(1):9–11

Current Flow Cytometry Methods for the Clinical Development of Immunomodulatory Biologics

Richard Wnek, Michelle Tseng and Dianna Wu

Abstract Flow cytometry is a highly versatile single cell analysis technology enabling multiparametric immune monitoring during clinical development. Cell data collected on morphology, activation state, and effector function from specific immune cell populations can be measured simultaneously in peripheral blood and/or tissue to determine drug efficacy. As such, flow cytometric laboratory testing is increasingly in demand to meet biomarker requirements for novel immunomodulatory biologics (IMBs) being evaluated in early and late stage clinical trials. In the context of clinical trial design, flow cytometric biomarkers may be used to assess safety, target engagement, and pharmacodynamics to enable clinical decision making. Flow cytometric applications such as immunophenotyping of TBNK immune cells is commonly employed as a safety biomarker to assess potential drug related immunotoxicities. IMB target engagement and pharmacodynamic effects on immune cell distribution and/or effector function can be integrated with traditional pharmacokinetic results to yield useful information relevant to dose selection, dosing intervals in addition to informing on mechanism of action. In all cases, a fit-for-purpose analytical validation is applied to flow cytometric assays prior to their clinical implementation to ensure confidence in biomarker measurements, while CRO placement of these assays offers a significant degree of standardization during clinical sample analysis to ensure high biomarker data quality.

Disclosure Statement Richard Wnek is an employee of Merck & Co., Inc. and holds stock in the company.

R. Wnek (✉)
Translational Molecular Biomarkers, Merck & Co., Inc., Rahway, USA
e-mail: richard_wnek@merck.com

M. Tseng
Amerimmune Immunology Laboratory, O & O Alpan, LLC, Fairfax, VA 22030, USA
e-mail: mtseng@oandoalpan.com

D. Wu
Oncology Translational Research, Janssen Research and Development, Spring House, USA
e-mail: dwu34@its.jnj.com

R. Weiner and M. Kelley (eds.), *Translating Molecular Biomarkers into Clinical Assays*, AAPS Advances in the Pharmaceutical Sciences Series 21,
DOI 10.1007/978-3-319-40793-7_13

Keywords

Immunomodulatory biologic Biotherapeutic entities that target the activation or suppression of host immune responses as a means of disease treatment.

Clinical development The clinical trials phase of drug development purposed with determining safety and efficacy of novel molecular entities.

Flow cytometry Fluorescence based technology capable of simultaneously resolving phenotype and effector function of cells present in complex biological matrices.

Immune monitoring Functional characterization and/or enumeration of immune cell subsets in peripheral blood and/or tissue.

Introduction

Immunomodulatory biologics (IMBs) are biotherapeutic entities that enhance or suppress the host immune response as a means of treating various cancers and complex immune disorders. With their targeted selectivity and superior clinical efficacy, IMBs represent an emerging therapeutic strategy and investment opportunity for disease intervention by many pharmaceutical and biotech companies. Novel biopharmaceutical IMBs entering into clinical development demand highly specialized cell-based assays to address biomarker requirements. As a result, flow cytometric laboratory testing is becoming increasingly important to the clinical evaluation of IMBs as it offers automated, high-throughput, multiparameter analysis of immune cell phenotype and effector function in complex biological fluids. Throughout this chapter we will focus on flow cytometric applications for immune monitoring in peripheral blood as a tool for biomarker analysis in the context of IMBs. Considerations for the validation and implementation of flow cytometric biomarker assays in global, multicenter clinical trials will also be emphasized.

Overview of Immunomodulatory Biologics

Immunotherapy via the administration of biologics targeting the modulation of immune responses has therapeutic utility in a broad range of clinical indications including diseases of immune dysfunction (i.e., Crohn's disease, rheumatoid arthritis), chronic infection, and cancer [1]. In general, IMBs have an attractive pharmacological profile including high on-target effects, a long half-life, and broad extracellular fluid biodistribution which make them attractive biopharmaceutical candidates for use in disease intervention [2]. IMBs can target specific immune cell subsets, such as ipilimumab, which blocks CTLA4 (CD152) on T-cells, or target soluble immune mediators, such as infliximab, which binds the cytokine TNF to block its interaction with TNFR [1, 3]. As such, IMBs are commonly grouped into three main categories based on their modes of action and include: recombinant human cytokines (i.e., IL-2, GM-CSF, IFN-α, β, γ), antibody-based triggering IMBs (i.e., alemtuzumab, rituximab), and blocking IMBs (i.e., pembrolizumab). Cytokine mimicry, cell depletion, activation, and suppression of immune cell effector responses are but a few key examples of how clinically administered IMBs exert their therapeutic effect [1].

Modulation of immune cell subsets using agonistic or antagonistic monoclonal antibodies (mAbs) is perceived to be a major opportunity for the treatment of various solid tumor and hematological malignancies [3]. In particular, a developing pipeline of therapeutic IMBs has emerged targeting immune checkpoint receptors on T-cells. Building on the success of ipilimumab, next generation blocking IMBs targeting T-cell co-inhibitory receptor PD-1 (CD279) have recently received FDA approval and those targeting one of its ligands PD-L1 (B7-H1) are currently in clinical development. To date, anti-PD-1 mAb clinical trial results show promising efficacy and safety profiles in multiple tumor types with the hope of incorporation into various treatment regimens for long-lasting clinical benefit [4, 5]. As such, this has set the stage for other promising T-cell immunomodulatory approaches to cancer treatment including the development of mAb agonists to co-stimulatory molecules OX-40, GITR, and CD40L as a means to augment the proliferative capacity, activation, and effector cytokine function of $CD4^+$ and $CD8^+$ T-cells during anti-tumor responses [2].

Flow Cytometry as a Biomarker Assay Platform to Support the Clinical Development of Immunomodulatory Biologics

Because IMBs have a broad range of clinical applications, targets, and modes of action it is important to incorporate sensitive and robust biomarker assays tailored to the specific biologic and its intended cellular target(s) throughout the clinical development process. Target engagement, predictive, and pharmacodynamic (PD) biomarkers are all vital to monitoring the ability of IMBs to modulate the cells and/or signaling pathways involved in the initiation and termination of effective immune responses. Furthermore, these biomarkers in the context of clinical trial design offer valuable information on dose selection, on-treatment dynamics of immune therapy, and clinical risk monitoring for exaggerated pharmacology and adverse reactions.

Flow cytometry is an ideal biomarker assay platform for the clinical evaluation of IMBs (Table 1). It is a high-throughput and information rich analytical technology that uses fluorescent dye conjugated reagents to simultaneously resolve

Table 1 Flow cytometric biomarker assays for the clinical development of immunomodulatory biologics

Flow cytometric assay	Assay description	Example application	Immunomodulatory biologic	Intended data use
Immunophenotyping	Characterization and/or enumeration of immune cell subsets using direct/indirect mAb staining against cell surface and/or intracellular antigens	TBNK analysis	Rituximab (ant-CD20 mAb)	Pharmacodynamic biomarker (7, 8)
		FoxP3+ Treg. analysis	Infliximab (anti-TNFα mAb)	Predictive biomarker (10)
Effector cell function	Cellular measurement of functional responsiveness using immunophenotyping, intracellular cytokine staining and/or phosphoflow methods to detect changes in protein expression, intracellular effector cytokine production and phosphorylation of signaling proteins	ICOS expression on CD4+ T-cells Ki67 expression on CD8 + T-cells	Ipilimumab (ant-CTLA4 mAb)	Pharmacodynamic biomarker (11)
		IL-17 cytokine producing $CD4^+$ T-cells (TH17)	Tremelimumab (anti-CTLA4 mAb)	Mechanism of action biomarker (12)
		Phosphoepitope signaling network analysis (ex. pSTATs, pERK pp38)	Tremelimumab (anti-CTLA4 mAb)	Pharmacodynamic biomarker (13)
Receptor occupancy (RO)	The measurement of drug targeted cell surface receptor sites pre- and post-drug administration using a combination of detection mAbs that recognize free (unbound), drug bound and/or total receptors	CD86 RO	Belatacept (anti-CTLA4 mAb)	Target Engagement (16)
		PD-1 RO	MDX-1106 (anti-PD1 mAb)	Target Engagement (5)
		CD33 RO	AVE9633 (anti-CD33 maytansine conjugate)	Target Engagement (15)

morphology, antigen expression, and effector function of individual cells present in complex biological matrices such as blood, bronchoalveolar lavage, and bone marrow aspirate. The technology works by hydrodynamically focusing thousands of fluorescent reagent(s) stained cells in suspension within a sample stream toward a laser-based interrogation point where beam steering optics direct emitted fluorescence from cells into photomultiplier tubes for conversion into digital signals by electronic components within the instrument. Digital signals are then converted into a standardized file format (.fcs) for visualization as histograms and/or bivariate dot plots using specialized software packages [6]. More advanced computational and visualization tools are also emerging for complex data sets. Most clinic facing cytometers to date are capable of measuring at minimum ≤10 biomarkers of relevant immune cell phenotype and/or effector function within a single patient sample. As such, flow cytometry is a highly versatile technology to support clinical biomarker analysis.

Immunophenotyping

Extensive immune cell characterization and/or enumeration can be achieved in a single patient sample using a combination of direct/indirect staining methods with fluorochrome-conjugated mAbs against cell surface and intracellular antigens. Clinical laboratory testing of major T-, B-, and natural killer (TBNK) immune cell subsets can be accomplished using commercially available lyophilized mAb kits with reference ranges established for major peripheral blood leukocyte populations in healthy volunteers. As such, TBNK immunophenotyping is routinely employed as a safety and/or PD biomarker in clinical trials. For IMBs, TBNK analysis has been used to enumerate $CD19^+$ B-cell counts pre- and post-treatment with rituximab, the anti-CD20 mAb, to determine treatment response and differentiate patient responders from non-responders [7, 8].

Immunophenotyping of major immune cell subsets can also be combined in parallel with other mAb to further differentiate cell subpopulations as treatment relevant biomarkers. A variety of phenotypic markers on T-cells including CD45 RA or RO, CCR7, CD25, CD127, FoxP3, HLA-DR are commonly multiplexed with major lineage markers to identify naïve, memory, regulatory, and activated T-cell subsets [9]. As an example, phenotypic analysis and enumeration of regulatory $CD4^+$ T-cells (Tregs) differentially expressing CD25/CD127 and intracellular FoxP3 is becoming an increasingly important consideration in the development of IMBs as their presence can impact the effect of immunotherapy on disease pathology. For instance, Treg modulation is hypothesized to be a possible mechanism of action for anti-TNFα agents in various autoimmune diseases such as inflammatory bowel disease (IBD). To this effect, increased frequencies of Tregs in peripheral blood were detected only in IBD clinical responders treated with infliximab suggesting peripheral Treg levels may correlate with clinical response [10].

Functional Assays

Flow cytometric biomarker assays that characterize cell-mediated immune responses combine standard immunophenotyping with a functional measurement of cellular physiology such as surface protein expression, intracellular cytokine production, or cell signaling induced phosphorylation of transcription factors and kinases [6]. Functional biomarker assays have diverse applications including PD assessment toward modeling PK/PD relationships, proof of pharmacology, and/or mechanism of action in clinical trials.

Modulation of cell surface or intracellular proteins associated with cell activation or suppression can be used to monitor PD effects following immunotherapy. Percent expression or median fluorescence intensity measurements converted into molecules of soluble fluorescence (MESF) using reference beads for standardization purposes may be used to measure longitudinal changes in protein expression in patient-derived samples during the course of a clinical trial. In regard to IMBs, CTLA4 blockade with ipilimumab stimulates anti-tumor immune responses [3]. As such, increased expression of the activation marker, ICOS, and the proliferation marker, Ki67, on circulating $CD4^+$ and $CD8^+$ T-cells have been reported as PD biomarkers in melanoma patients following treatment with ipilimumab [11]. Other potential markers of T-cell activation may include HLA-DR, CD27, OX40, 4-1BB, and CD40L. In contrast, increases in BTLA, PD-1, LAG3, and TIM-3 markers may signify functional T-cell exhaustion and/or inhibition [3, 9].

De novo synthesis of intracellular effector cytokines or phosphorylation of cell signaling proteins can also be used as a measure of functional responsiveness to immunotherapy. This is accomplished by combining cell surface immunophenotyping with intracellular cytokine staining (ICS) or phosphoflow methods using fixatives and permeabilization agents to gain access to cell cytoplasmic and/or nuclear compartments prior to mAb staining. As such, ex vivo assessment of T-cell functional competence by ICS and phosphoflow analysis has the potential to identify treatment relevant biomarkers. For example, intracellular cytokine staining in melanoma patients following CTLA4 blockade with the mAb tremelimumab revealed increases in IL-17 cytokine producing $CD4^+$ cells following ex vivo stimulation [12]. These TH17 cells were preferentially increased in patients that developed clinically relevant autoimmune toxicities after one round of immunotherapy, highlighting mechanistic and dose limitations to the clinically administered mAb. Similarly, phosphoflow analysis in tremelimumab treated melanoma patients showed alterations in $CD4^+$ and $CD8^+$ T-cell signaling pathways. Increases in pSTAT1, pSTAT3, pp38 along with decreases in pLck, pERK1/2, pSTAT5, demonstrated PD effects consistent with direct inhibition of T-cell signaling downstream of the T-cell receptor (TCR) complex; the suspected mechanism of action for CTLA4 blocking mAbs [13].

Receptor Occupancy

Flow cytometric assessment of cell surface receptors and measurement of receptor occupancy (RO) pre- and post-IMB administration provides valuable information to confirm target engagement, to inform on dosing intervals, and to develop PK/PD relationships in early clinical trials. Flow cytometric RO assays have been previously reported for the pre-clinical development of small molecule antagonists in peripheral blood but the application is also amenable to IMB clinical development [14]. RO can be monitored on any leukocyte subset in peripheral blood if the target receptor is normally expressed. Receptor dynamics should be considered as transient, low expressing, and/or internalizing cell surface receptors present challenges during RO assay development. RO may also be assessed in certain cases at the desired site of drug action. As example, RO on malignant myeloblasts in bone marrow aspirates collected from patients with refractory acute myeloid leukemia has been previously reported in phase I studies with anti-CD33 immunoconjugates [15].

Flow cytometric RO assays are esoteric in design and execution but commonly employ fluorochrome-conjugated mAbs that differentially recognize epitopes to semi-quantitatively measure unbound (free), IMB occupied and/or total target receptors pre- and post-drug administration. In some instances, antibody binding bead standards are additionally used to quantitate the antibody binding capacity (ABC) of the detection mAbs to their receptors as a correlate representation of the absolute number of target receptors expressed on the cell surface. When implemented in a clinical trial setting, data from flow cytometric RO assays is compared longitudinally, often reporting RO relative to pre-dose baseline levels following administration of the IMB.

As a clinical development example, two anti-CD86 mAb were employed to determine RO for belatacept in clinical trials, a second generation CTLA4 Ig (LEA29Y) which binds CD86 expressed on antigen presenting cells [16]. The primary anti-CD86 mAb chosen had a high binding capacity for its target receptor and was able to detect low levels of belatacept RO without displacing it which in general are desirable characteristics when choosing a detection mAb to assess target engagement. In contrast, a non-competitive anti-CD86 mAb clone was also identified; one that did not interfere with belatacept and as such was used to calculate the total number of target receptor molecules. Alternatively, saturating amounts of the biotherapeutic IMB under clinical evaluation may be employed to quantitate RO using indirect mAb staining methods. To this effect, saturation with an anti-PD-1 mAb (MDX-1106) under clinical evaluation in a Phase I study of patients with refractory solid tumors was used to quantitate occupied PD-1 receptor sites and the total number of available PD-1 binding sites [5]. Specifically, the ratio of change in mean fluorescence intensity (rather than percentage positive events) of $CD3^+$ lymphocytes pre-incubated ex vivo with saturating amounts of an isotype control antibody (indicating in vivo binding) or anti-PD-1 (to detect total available binding sites) was used to estimate PD-1 RO [5].

Assay Validation of Flow Cytometric Biomarker Assays

Assay validation of flow cytometric biomarker assays is a complex process due to the limited post-collection stability of clinical specimens, a lack of quality control (QC) reference standards, and technical variations between analytical laboratories [17]. Because flow cytometric applications are varied, validation parameters will be somewhat different across assay types. As such, flow cytometric biomarkers are validated using the fit-for-purpose paradigm which offers flexibility in validation requirements to meet the intended use of the data generated in a resource-effective fashion [18]. At minimum, analytical validation using fresh specimens will include an assessment of post-collection specimen stability, intra-/inter-assay precision and intra-subject variability testing in order to differentiate the effects of post-collection specimen handling, sample processing (often involving manual procedures), and inherent biological variability from the IMB mediated pharmacological effects on clinical specimens. In some instances, commercial disease specimens may also need to be procured from biorepository vendors prior to analytical validation.

When fresh specimens are desired, post-collection stability is first established to determine biomarker stability limitations from the time of specimen collection to flow cytometric data acquisition. Once established, standardized instruction procedures for specimen collection, handling and shipping conditions are to be provided to all participating clinical study sites to ensure clinical specimen integrity within the assay stability window. Next, intra-assay (within-run) and inter-assay (between-run) precision is evaluated to measure variability in assay performance. For immune cell subsets that are at least 10 % of the parental cell population, intra- and inter- assay precision with a coefficient of variability (CV) less than 20 % is generally considered acceptable. In instances where rare cell populations (<10 %) such as Tregs are to be measured, a CV up to 30 % may be acceptable but acceptance criteria should always be defined in context of intended clinical data use [19]. To further limit variability, all participating laboratory analysts must be well-trained and demonstrate competency before participating in assay validation. Finally, an assessment of intra-subject variability (within donors) is conducted in order to identify potential diurnal variations. Once complete, final results from validation experiments must be documented and reviewed to ensure adequate confidence in the measurements.

Assay Implementation in Clinical Trials

In order to ensure consistent flow cytometric assay performance and high-quality biomarker data across all regional laboratories over the course of a global, multi-center clinical trial implementation of appropriate QC materials during clinical sample analysis and the effective management of analytical variables using a contract research organization (CRO) is vital to successful assay implementation.

QC materials are used to monitor sample processing and flow cytometric assay performance enabling analysts to capture analytical errors in real time, to resolve technical issues and perform reanalysis within specimen stability limits as needed. QC materials should mimic the clinical specimen type and express the biomarkers of interest in the expected target range. In some instances, commercially available preserved blood specimens with limited phenotype and stability over several weeks can be used to monitor clinical sample analysis. Alternatively, frozen peripheral blood mononuclear cells and/or cell lines which express biomarkers of interest may also be used as QC materials. Once a QC material is identified, the laboratory establishes a lot-specific QC acceptance range to qualify analytical runs by evaluating the 95 % confidence interval from 10 to 20 analyses. If lot-specific QC data falls out of the acceptance range during clinical sample analysis, biomarker assay results are to be reevaluated and, if necessary, the laboratory testing is to be repeated [17].

To meet the increasing demands for flow cytometric biomarkers in multicenter clinical trials, the placement of clinical sample analysis at CROs offers a significant degree of standardization. This is attributed to the CRO use of identical flow cytometry instruments, SOP-driven methods, and centralized peer review of data quality; all serving to minimize biomarker assay variation in the clinical data set [20]. When evaluating a CRO, a detailed review of scientific expertise and laboratory personnel balanced with an infrastructure assessment for supporting early and late stage global clinical trials is highly recommended. The scientific proficiency of a CRO should initially be assessed by means of a pilot study using a well-established biomarker assay developed by the sponsor group. Of note, routine communication between the sponsor and the CRO laboratory is vital to successful assay implementation [20].

Summary

Therapeutic IMBs represent an emerging strategy for disease intervention; in particular, for the treatment of various cancers using mAbs to stimulate host anti-tumor immune responses. As novel IMBs enter into clinical trials, flow cytometric immune monitoring is increasingly in demand to meet biomarker requirements for the clinical development of IMBs. Flow cytometric biomarker assays incorporating immunophenotyping, analyses of effector function and receptor occupancy are broadly utilized to confirm target engagement, explore PK/PD correlations, interogate mechanism of action, and predict treatment response to immunotherapy; all of which provide valuable information for guiding clinical decisions and supporting IMB filings with regulatory agencies. Like most cell-based assays validation and implementation of flow cytometric biomarker assays in global, multicenter clinical trials remains a complex process. A fit-for-purpose approach to assay validation ensures adequate confidence in biomarker measurements while assay implementation using CRO laboratories offers a significant degree of standardization for the effective management of analytical variables encountered during clinical sample analysis.

Summary Box

- Immunomodulatory biologics (IMBs) represent an emerging strategy for disease intervention by pharmaceutical and biotech companies; in particular for the treatment of complex immune disorders, (i.e., Crohn's disease, rheumatoid arthritis), chronic infection, and cancer.
- Flow cytometric laboratory testing is becoming increasingly important to the clinical evaluation of IMBs as it offers automated, high-throughput, multiparameter analysis of immune cell phenotype, and effector function in complex biological fluids.
- Flow cytometric biomarker assays are commonly used to confirm target engagement, explore PK/PD correlations, interrogate mechanism of action during clinical trials.
- Flow cytometric biomarker assays are validated using the fit-for-purpose paradigm which offers flexibility in validation requirements to meet the intended use of the data generated in a resource-effective fashion.
- Minimum requirements for the analytical validation of flow cytometric assays using freshly collected blood specimens include an assessment of post-collection specimen stability, intra-/inter-assay precision and intra-subject variability testing.
- The identification and use of appropriate QC materials during clinical sample analysis and the effective management of analytical variables using a contract research organization (CRO) help ensure consistent flow cytometric assay performance and high quality biomarker data over the course of a global, multicenter clinical trial.

References

1. Sathish JG, Sethu S, Bielsky MC et al (2013) Challenges and approaches for the development of safer immunomodulatory biologics. Nat Rev Drug Discov 12(4):306–324
2. Melero I, Grimaldi AM, Perez-Gracia JL, Ascierto PA (2013) Clinical development of immunostimulatory monoclonal antibodies and opportunities for combination. Clin Cancer Res 19(5):997–1008
3. Ascierto PA, Kalos M, Schaer DA, Callahan MK, Wolchok JD (2013) Biomarkers for immunostimulatory monoclonal antibodies in combination strategies for melanoma and other tumor types. Clin Cancer Res 19(5):1009–1020
4. Hamid O, Robert C, Daud A et al (2013) Safety and tumor responses with lambrolizumab (anti-PD-1) in melanoma. N Engl J Med 369(2):134–144
5. Brahmer JR, Drake CG, Wollner I et al (2010) Phase I study of single-agent anti-programmed death-1 (MDX-1106) in refractory solid tumors: safety, clinical activity, pharmacodynamics, and immunologic correlates. J Clin Oncol 28(19):3167–3175
6. Barnard RM (2012) Flow cytometry: a flexible tool for biomarker research. Bioanalysis 4(20):2471–2483

7. Fluge O, Bruland O, Risa K et al (2011) Benefit from B-lymphocyte depletion using the anti-CD20 antibody rituximab in chronic fatigue syndrome. A double-blind and placebo-controlled study. PLoS ONE 6(10):e26358
8. Sellam J, Hendel-Chavez H, Rouanet S et al (2011) B cell activation biomarkers as predictive factors for the response to rituximab in rheumatoid arthritis: a six-month, national, multicenter, open-label study. Arthritis Rheum 63(4):933–938
9. Lacey SF, Kalos M (2013) Biomarkers in T-cell therapy clinical trials. Cytotherapy 15 (6):632–640
10. Guidi L, Felice C, Procoli A et al (2013) FOXP3(+) T regulatory cell modifications in inflammatory bowel disease patients treated with anti-TNFalpha agents. Biomed Res Int 2013:286368
11. Wang W, Yu D, Sarnaik AA et al (2012) Biomarkers on melanoma patient T cells associated with ipilimumab treatment. J Transl Med 10:146
12. Von Euw E, Chodon T, Attar N et al (2009) CTLA4 blockade increases Th17 cells in patients with metastatic melanoma. J Transl Med 7:35
13. Comin-Anduix B, Sazegar H, Chodon T et al (2010) Modulation of cell signaling networks after CTLA4 blockade in patients with metastatic melanoma. PLoS ONE 5(9):e12711
14. Koo GC, Shah K, Ding GJ et al (2003) A small molecule very late antigen-4 antagonist can inhibit ovalbumin-induced lung inflammation. Am J Respir Crit Care Med 167(10): 1400–1409
15. Lapusan S, Vidriales MB, Thomas X et al (2012) Phase I studies of AVE9633, an anti-CD33 antibody-maytansinoid conjugate, in adult patients with relapsed/refractory acute myeloid leukemia. Invest New Drugs 30(3):1121–1131
16. Latek R, Fleener C, Lamian V et al (2009) Assessment of belatacept-mediated costimulation blockade through evaluation of CD80/86-receptor saturation. Transplantation 87(6):926–933
17. Wu DY, Patti-Diaz L, Hill CG (2010) Development and validation of flow cytometry methods for pharmacodynamic clinical biomarkers. Bioanalysis 2(9):1617–1626
18. Lee JW, Devanarayan V, Barrett YC et al (2006) Fit-for-purpose method development and validation for successful biomarker measurement. Pharm Res 23(2):312–328
19. O'hara DM, Xu Y, Liang Z, Reddy MP, Wu DY, Litwin V (2011) Recommendations for the validation of flow cytometric testing during drug development: II assays. J Immunol Methods 363(2):120–134
20. Hill CG, Mccloskey TW (2011) Flow cytometry outsourcing. Eur Biopharm Rev (Autumn 2011)

Key Mass Spectrometry Techniques Used in Clinical Biomarker Research

Mingxiang Lin

Abstract Mass spectrometry has been playing an increasingly important role in various aspects of biomarker research, ranging from discovery of novel biomarkers to quantitative measurement of known biomarkers for clinical applications. Following a brief overview of multiple reaction monitoring (MRM)-based LC/MS approach, the gold standard for quantitative mass spectrometry, this chapter provides an overview of several emerging mass spectrometry techniques benefit from recent advances in mass spectrometry instrumentation and related technologies. These new techniques enable researchers to develop sensitive, specific, robust, and higher throughput biomarker assays for novel clinical applications. Multiple reaction monitoring cubed (MRM^3) technique is capable of measuring analytes in complex matrices without extensive sample pretreatment. High-pressure, high-resolution separations with intelligent selection and multiplexing (PRISM) technique greatly increase the efficiency of complex sample analysis in a highly automatic fashion. High-resolution mass spectrometry (HRMS) enables the extraction of analyte information from complicated matrices with minimal efforts in method development. Parallel reaction monitoring (PRM) offers an enhanced MRM approach with better tolerance to background interference and the potential of achieving better detection limits. Microfluidic LC/MS utilizes low-flow LC to boost sensitivity tremendously and enables the detection of extremely low-abundance analytes. Stable isotope dilution mass spectrometry (SID-MS) enables accurate measurement of protein and peptide in biological matrices. These new technologies have changed the landscape of MS usage in clinical biomarker field and will continue to bring positive impact as more advanced tools become available.

Keywords Emerging mass spectrometry techniques · MRM^3 · PRISM · HRMS · PRM · Microfluidic LC/MS · SID-MS

M. Lin (✉)
Analytical Research & Development, Merck & Co., 126 E. Lincoln Ave, Rahway 07065, NJ, USA
e-mail: mingxiang.lin@merck.com

R. Weiner and M. Kelley (eds.), *Translating Molecular Biomarkers into Clinical Assays*, AAPS Advances in the Pharmaceutical Sciences Series 21,
DOI 10.1007/978-3-319-40793-7_14

Overview

Development of a biomarker strategy can be viewed as a multistage process starting from early discovery, followed by verification and validation, and ultimately biomarker measurement in clinical samples. Mass spectrometry has been playing an increasingly important role in every aspect of biomarker research [1]. Depending on the stage of the biomarker development, mass spectrometry techniques involved have different characteristics. At the discovery stage, high-resolution, full scan MS technique coupled with complex fractionation technique is ideal for identifying novel biomarkers with biomedical implications. At the clinical application stage, sensitive quantitative MS technique with high-throughput capability is required to meet the need for accurate and robust biomarker measurement [2]. While the previous two chapters have provided an overview on the MS techniques used for small molecule and large molecule biomarker measurement applications, respectively, this chapter provides a brief review on new development in MS and related technologies that will have potential positive impact on either expanding the utility of MS or improving the performance of MS-based platforms in clinical biomarker applications.

MRM—The Gold Standard

The MS technique suitable for clinical biomarker measurement must have the following characteristics: high sensitive, high specificity, broad dynamic range, and fast acquisition speed to couple with high-throughput liquid chromatography separation. To date, among available MS-based quantitative techniques, multiple reaction monitoring (MRM), also known as selected reaction monitoring (SRM) carried out in triple quadrupole instruments, has been the core technique as it possesses the necessary features to perform reliable quantitation measurement [3]. In a typical LC-MRM approach, an analyte of interest is ionized to form a precursor ion after LC separation via a soft ionization source, most commonly electrospray. The precursor ion is then mass isolated in the first quadrupole Q1 and subjected to collision-induced dissociation (CID) in the present of collision gas at elevated collision energy in a collision cell Q2. A product ion resulting from CID fragmentation was then selected and monitored at the third quadrupole Q3 to generate MRM signal for quantitation. MRM has been traditionally used in small molecule applications, such as pharmacokinetics and ADME quantitation studies, but it has recently expanded into peptide and protein quantitation with the introduction of extended mass range triple quadrupole instruments. As clinical biomarker research enters the arena of quantifying low-abundance proteins/peptides in complex matrices (e.g., plasma, serum), there has been an ever-increasing demand for higher sensitivity and better specificity. Although MRM is an inherently sensitive MS technique, it is a low mass resolution-based approach and is prone to background interference, especially when dealing with complex matrices. With the advances in instrumentation, researchers have developed new quantitation approaches to utilize new capabilities offered by novel MS and LC technologies. The following sections will provide a broad overview of several promising emerging techniques available to the researcher in biomarker quantitation field.

New MS Techniques

MRM^3

When using MRM to measure analytes in complex matrices, extensive sample fractionation or target analyte enrichment is often required to remove interference as much as possible, which usually leads to increase assay complexity and lower sample throughput, and thus hinders the implementation of such assays in clinical laboratories mostly dealing with routing analysis. A technique that utilizes MS^3 capability of a hybrid triple quadruple/linear ion trap mass spectrometer has been developed to address this challenge. This technique is termed multiple reaction monitoring cubed, MRM^3 [4, 5]. Compared to conventional MRM, in which quantitation is achieved through a two-stage MS fragmentation, MRM^3 can be viewed as two MRM processes used in tandem with one additional CID fragmentation to achieve higher specificity. MRM^3 is typically conducted in a hybrid linear ion trap instrument, in which the third quadrupole is replaced by a linear ion trap and thus enables trapping and further fragmenting the primary MS^2 product ions. After MS^3 fragmentation, an MRM^3 chromatogram is then reconstructed by monitoring the intensity of a MS^3 product ion in the linear ion trap. The additional fragmentation step can drastically reduce or even completely remove inferences observed in conventional MRM and achieve a better signal-to-noise ratio, which might lead to a better lowest limit of quantitation (LLOQ) without the need of extensive sample pretreatment. It should be noted that MRM^3 might decrease sensitivity due to ion signal loss during the additional MS/MS. Also, this approach cannot reduce matrix effect due to ion suppression in ion source. A pre-MS-analysis separation/fractionation is necessary to reduce such matrix effect.

PRISM

Another triple quadrupole-based technique is called high-pressure, high-resolution separations with intelligent selection and multiplexing (PRISM) [6, 7]. PRISM basically is a technique that fractionation peptides from a complex sample using high-resolution reverse phase LC, while monitoring peptides in each fraction via multiplexing MRM. The implement of PRISM strategy is a two-step process. The first step is to fractionate digested samples along with stable isotope labeled internal standards via reserve-phase capillary LC separation using basic pH mobile phase. Unlike commonly used strong cation exchange (SCX) crude fractionation, the reserve-phase fractionation provided a higher resolution separation and the fraction is compatible with downstream MS analysis. During the LC fractionation, a small stream of the LC flow is split off and continuously monitored by online MRM of labeled internal standards to identify fractions containing the peptides of interest. In the second step, the peptides in selected fractions are quantified by nanoflow LC-MRM using acidic mobile phase. The different pH in mobile phase used between the first and the second dimension LC offers a partial orthogonality and improves fractionation efficiency. PRISM can greatly reduce sample complexity and matrix effect caused by ion suppression via the high efficient reverse phase LC fractionation process. The PRISM-MRM platform has been demonstrated to be a sensitive platform capable of quantifying low ng/mL proteins with small volume of plasma sample (<20 μL). Coupled with front-end immunoaffinity depletion, the quantitation limit can be extended into 50–100 pg/mL range. A possible disadvantage of the PRISM compared with direct LC/MS analysis is reduced throughput as the need for two-stage LC separation and multiple fractionation analysis. A mitigation strategy is to combine several non-interfering fractions as the content of each fraction has been identified during the first step.

HRMS

High-resolution mass spectrometry (HRMS) is traditionally associated with qualitative analysis and not suitable for quantitative analysis, as the sensitivity and dynamic range offered by a HRMS instrument are typically inferior to a triple quadrupole instrument. With the recent advance in instrumentation, newer HRMS instruments, such time-of-flight (TOF) and Orbitrap, have greatly improved sensitivity and dynamic range, and have been demonstrated to be capable of conducting reliable and robust quantitative analysis [8, 9]. In contrast to MRM where quantitation is achieved through MS/MS fragmentation, HRMS quantitation is based on intact precursor ions acquired under full scan MS mode. There are several advantages associated with this approach. (1) Unlike MRM approach, where a preset of mass transition optimized for each analyte is monitored, a full scan approach monitors all ions present in the ion source. The ability to quantify ion of interest relies on high-resolution accurate mass measurement, which enables the construction of XIC on the analyte's theoretical m/z with a narrow window and thus achieve high specificity. This approach is especially advantageous for HRMS to obtain quantitative information when dealing with complex samples, where MRM might suffer from interferences. (2) Full scan acquisition offers a complete picture for the analyte, such as adducts and charges states, and simplifies assay method development and troubleshooting. (3) An MRM quantitation method results in the best sensitivity when a dominant MS/MS transition can be identified. However, in a scenario of poor fragmentation (the precursor ion is resistant to CID) or extensive fragmentation (a good MRM cannot be identified as MS/MS signal spreads across multiple pathways), full scan-based HRMS quantitation would be a better alternative to MRM. Despite the advantages of HRMS quantitation discussed here, there are two main factors so far limit a wider adaption of HRMS in quantitative applications, especially for clinical use. One is that there still is a gap in raw sensitivity between HRMS and triple quadrupole-based MRM approach for most applications, especially for small molecules. The other is that HRMS instruments generally cost more to acquire and require more skills to operate and maintain, while triple quadruple instrument is relatively easy to deploy in a clinical laboratory setting. Nevertheless, as instrument vendors continue to improve the sensitivity of HRMS and aim for a more user-friendly operation, HRMS is gradually gaining popularity and would be an important tool in biomarker quantitation field in the foreseeable future.

PRM

Another new MS quantitation technology termed parallel reaction monitoring (PRM), which takes advantage of newer hybrid quadrupole equipped high-resolution hybrid mass spectrometers, has emerged recently [10]. Although also based on HRMS, the operation of PRM actually more closely resembles traditional MRM quantitation and is different from the full scan approach outlined in the previous section. Performed in a quadruple time-of-flight (Q-TOF) or a quadruple Orbitrap instrument, PRM can be envisioned as an extended MRM, where the third quadrupole of a triple quadrupole is substituted by a high-resolution mass analyzer to permit parallel detection of all product ions from a targeted precursor with high mass accuracy. PRM has several potential advantages over MRM. (1) PRM approach provides high degree of specificity since all MS/MS product ions are available to confirm the identity of target molecules. (2) PRM is more tolerant to background interference since product ions are acquired with good mass accuracy under high-resolution mode. The high-resolution MRM chromatogram generated in PRM is an extract ion chromatogram with a narrow m/z window of a selected product ion, thus effectively remove many interfering background ions that would be problematic in a conventional MRM operation. (3) Since all product ions are monitored, a PRM method would require less effort in method optimization. It should be recognized that PRM is inherently a less sensitive approach compared with MRM due to the lower duty cycle of scanning mode in PRM. It has been shown that PRM can achieve similar performance benchmarks in terms of dynamic range and linearity, but to a lesser extent of precision. However, the high selectivity and specificity enable PRM to overcome limitations in speed and sensitivity and might provide lower quantitation limits with minimal upfront efforts for method development.

Microflow LC and Chip-Based Micro Fluidic Devices

Besides the new MS quantitation techniques discussed above, improvement in LC technology has also greatly expanded the capability of LC-MS/MS-based quantitation method. One of the areas in LC development has generated particular interests is microflow LC technique [11, 12]. It has been recognized that operating LC at a lower flow rate can increase sensitivity as the electrospray ionization efficiency increases as the flow rate decreases. Nanoflow LC with less than 1 μL/min flow rate has been routinely used in proteomics research to aim for ultimate sensitivity. However, nanoflow is notoriously known for its low speed and difficulty to use. It also lacks the robustness and high-throughput characteristics desired in bioanalytical analysis. Microflow LC operating at 1–50 μL/min flow rate has been developed as a low-flow LC approach effectively balancing sensitivity and throughput. It has been demonstrated that microflow LC can be used to develop a validated LC/MS quantitation assay meeting the stringent requirements of regulated bioanalysis [12]. The head-to-head comparison between microflow LC and conventional flow LC in this study reveals that microflow LC-based assay has improved detection limits through high MS signal (more than tenfold improvement in signal-to-noise ratio), while achieving similar performance benchmarks in accuracy, precision, matrix effect, and general robustness. Additional benefits offered by microflow LC include reduced MS source contamination and solvent consumption.

Another hot area in microflow LC development is chip-based microfluidic devices, which further improve the robustness and ease of use of microflow LC. A typical chip-based microfluidic device integrates capillary LC column with electrospray emitter in a chip format. This configuration greatly simplifies the connections between capillary tubing and results in an easy-to-change user interface. Zhu et al. have developed and validated an Agilent microfluidic chip-based LC-MS/MS method to simultaneously quantify several abused drugs and their metabolites in human hair [13]. Broccardo et al. have used Trizaic nanoTile, a similar technology offered by Waters, to develop and validate a multiplexed LC/MS assay for five steroid hormones in human serum [14]. This Trizaic platform achieves an increase of 100–300-fold on-column sensitivity while decreasing solvent consumption by 150-folds. The clinical performance of this Trizaic-based microflow LC/MS has been further compared with antibody-based immunoassay, and a greatly improved sensitivity and specificity is reported with the Trizaic approach. Overall, with the help of microfluidic devices, bioanalytical scientists now have access to the great sensitivity offered by microflow LC with relatively ease. As the technology continues to evolve, microflow LC would become the method of choice for bioanalytical applications demanding high sensitivity.

Stable Isotope Dilution MS

Clinical assays typically deal with complex biological sample matrices and require various degree of sample treatment. To overcome the detrimental matrix effect and to compensate the loss of analyte signal during sample manipulation, internal standards are often required to obtain accurate measurement. One of most powerful techniques is stable isotope dilution mass spectrometry (SID-MS), in which stable isotope labeled analogs of target analytes are used as internal standards. Stable isotope label molecules possess the same chemical and physical properties as their unlabeled counterparts, and exhibit almost identical behaviors during sample pre-treatment, HPLC separation, MS ionization, and fragmentation. SID coupled with MRM has long been used for small molecule quantitation in clinical chemistry [15]. In a typical small molecule SID-MRM method, known amount of standard isotope labeled standards are added to sample matrix before any pretreatment and analysis. The MS analysis is conducted with two MRM transitions for each analyte, one for the unlabeled analyte and one for the labeled standard incorporating the mass shift originated from heavy isotope labeling. The quantitative information is derived from the chromatographic peak ratio of an analyte to its labeled internal standard obtained from their MRM chromatograms.

Using the same principle, SID-MRM method has also been applied to protein quantitation in plasma [16]. The quantitation of protein in such studies is typically accomplished through a unique surrogate peptide fragment of target proteins produced from enzymatic digestion (e.g., trypsin digestion). The stable isotope labeled proteins will be the ideal internal standards, however such standard could be difficult to obtain. In addition, if an antibody is used to enrich target protein, the difference in posttranslational modifications and protein folding between the native protein and the synthetic standard might result in discrepancy in immunoaffinity enrichment efficiency. A more practice approach will be using stable isotope labeled version of the surrogate peptide, which is added to sample matrix after protein digestion. The peptide standard is then used to compensate any loss and matrix effect in the downstream analysis at peptide level. The key to such a strategy is to ensure the complete and reproducible digestion of target proteins. A good example of this approach is the stable isotope standards with capture by anti-peptide antibodies (SISCAPA) [17, 18, 19]. In a typical SISCAPA experiment, the whole plasma is subjected to trypsin digestion to completely break down all proteins in a sample. Then stable isotope labeled peptide standards are added, followed by anti-peptide antibodies enrichment of surrogate peptides. Similar to small molecule quantitation, subsequent LC-MRM analysis after SISCAPA approach provides peptide separation and quantitation derived from the analyte-to-internal-standard ratio.

The advantages offered by SID are straightforward—isotope labeled analogs can be served as the ideal internal standards in most cases. The quantitation via SID-MS based on relative abundance between the analyte and the known standard, thus signal variation caused by analyte loss during sample manipulation, matrix effect and instrument drift can be corrected under most circumstances. There are also a few limitations associated with SID-MS including limited availability of isotope labeled standard, the need for high purity standard, and possible mass overlap between an analyte's natural isotopic distribution and its labeled standard.

Summary

In the past two decades, LC-MRM-based quantitation has grown into a core technique in quantitative analysis. Newer technologies, such as high-resolution quantitation, multistage fragmentation, microflow LC, and stable isotope dilution mass spectrometry, have emerged as promising techniques and expanded the tools available to conduct biomarker analysis. As instrument technology continue to advance, the landscape of MS usage in clinical biomarker field will continue to evolve from a single MRM-based technique to a collection of various MS techniques available to perform bioanalysis once unachievable not long ago.

References

1. Smith RD (2012) Mass spectrometry in biomarker applications: from untargeted discovery to targeted verification, and implications for platform convergence and clinical application. Clin Chem 58:528–530
2. Grebe SKG, Singh RJ (2011) LC-MS/MS in the clinical laboratory-where to from here? Clin Biochem Rev 21:5–31
3. Kitteringham NR, Jenkins RE, Lane CS, Elliott VL, Park BK (2009) Multiple reaction monitoring for quantitative biomarker analysis in proteomics and metabolomics. J Chromatogr B 877:1229–1239
4. Fortin T, Salvador A, Charrier JP, Lenz C, Bettsworth F, Lacoux X, Choquet-Kastylevsky G, Lemonine J (2009) Multiple reaction monitoring cubed for protein quantification at low nanogram/milliliter level in nondepleted human serum. Anal Chem 81:9343–9352
5. Jaffuel A, Lemonine J, Aubert C, Simon R, Leonard J-F, Gautier J-C, Pasquier O, Salvador A (2013) Optimization of liquid chromatography-multiple reaction monitoring cubed mass spectrometry assay for protein quantification: application to aquaporin-2 water channel in human urine. J Chromatogr A 1301:122–130
6. Shi T, Fillmore TL, Sun X, Zhao R, Schepmoes AA, Hossain M. Xie F, Wu S, Kim J-s, Jone N. Moore RJ, Pasa-Tolic L, Kagan J, Rodland KD, Liu T, Tang K, Camp DG, Smith RD, Qian W-J (2012) Antibody-free, targeted mass-spectrometric approach for quantification of protein at low pictogram per milliliter levels in human plasma/serum. PNAS 190: 15395–15400

7. Shi T, Sun X, Gao Y, Fillmore TL, Schepmoes AA, Zhao R, He J, Moore RJ, Kagan J, Rodland KD, Liu T, Liu AY, Smith RD, Tang K, Camp DG, Qian W-J (2013) Targeted quantification of low ng/mL level proteins in human serum without immunoaffinity depletion. J Proteome Res 12:3353–3361
8. Rochat B, Kottelat E, McMullen J. The future key role of LC-high-resolution-MS analyses in clinical laboratories: a focus on quantitation. Bioanalysis 4:2939–2958
9. Goodenough A (2011) High resolution mass spectrometry approaches for the quantification of proteins using stable isotopically labeled peptide, 21–30 May/June 2011
10. Peterson AC, Russell JD, Bailey DJ, Westphall MS, Coon JJ (2012) Parallel reaction monitoring for high resolution and high mass accuracy quantitative, targeted proteomics. Mol Cell Proteomics 11:1475–1488
11. Arnold DW, Needhma SR (2013) Micro-LC-MS/MS: the future of bioanalysis. Bioanalysis 5(11):1329–1331
12. Christianson CC, Johnson CJL, Needham SR (2013) The advantages of microflow LC-MS/MS compared with conventional HPLC-MS/MS for the analysis of methotrexate from human plasma. Bioanalysis 5(11):1387–1396
13. Zhu KY, Leung KW, Ting AKL, Wong ZCF, Ng WYY, Choi RCY, Dong TTX, Wang T, Lau DTW, Tsim KWT (2012) Microfluidic chip based nano liquid chromatography coupled to tandem mass spectrometry for the determination of abused drugs and metabolites in human hair. Anal Bioanal Chem 402:2805–2815
14. Broccardo CJ, Schauer KL, Kohrt WM, Schwartz RS, Murphy JP, Prenni JE (2013) Multiplexed analysis of steroid hormones in human serum using novel microflow tile technology and LC-MS/MS. J Chromatogr B 934:16–21
15. Lee MS, Kerns EH (1999) LC/MS applications in drug development. Mass Spectrom Rev 18:187–279
16. Keshishian H, Addona T, Burgess M, Kuhn E, Carr SA (2010) Quantitative, multiplexed assays for low abundance proteins in plasma by targeted mass spectrometry and stable isotope dilution. Mol Cell Proteomics 6:2212–2229
17. Kuhn E, Addona T, Keshishian H, Burgass M, Mani DR, Lee RT, Sabatine MS, Gerszten RE, Carr SA (2009) Developing multiplexed assays for troponin I and interleukin-33 in plasma by peptide immunoaffinity enrichment and targeted mass spectrometry. Clin Chem 55:1108–1117
18. Whiteaker JR, Zhao L, Anderson L, Paulovich AG (2010) An automated and multiplexed method for high throughput peptide immunoaffinity enrichment and multiple reaction monitoring mass spectrometry-based quantification of protein biomarkers. Mol Cell Proteomics 9:184–196
19. Kuhn E, Whiteaker JR, Mani DR, Jackson AM, Zhao L, Pope ME, Smith E, Rivera KD, Anderson NL, Skates SJ, Pearson TW, Paulovich AG, Carr SA (2012) Interlaboratory evaluation of automated, multiplexed peptide immunoaffinity enrichment coupled to multiple reaction monitoring mass spectrometry for quantifying proteins in plasma. Mol Cell Proteomics 11:1–14

Clinical Genomic Biomarker Assay Development: Technologies and Issues

Ken C.N. Chang and Matthew J. Marton

Abstract Clinical assay development is quite different from preclinical assay development, since a clinical assay requires special considerations to make sure a biomarker discovered preclinically will be translatable into clinical settings. Once the assay is developed, its pre-analytical and analytical validity will be confirmed through a series of analytical validation experiments that will document its performance characteristics. One challenge frequently encountered during the clinical sample testing is that many samples fail to meet minimum sample requirements, either due to the low quality or due to insufficient quantity. Special considerations are often needed for selecting biomarker assay platforms depending on the type of tissues used in the clinical trials. In this chapter, we provided a brief overview of most frequently encountered clinical genomic assays and dissected many of them to show their practical application in clinic based on our experiences.

Keywords FFPE (formalin fixed paraffin embedded) · Gene signature · Clinical genomics · Analytical validation · Next generation sequencing

Introduction and Clinical Biomarker Assay Development Process

Congratulations: Drug candidate XYZ-123 is ready for its first human clinical trial. Is the drug candidate hitting its intended target? What is the mechanism of action of the drug candidate? What is the patient population most likely to respond to the drug candidate? Scientists rely on pharmacodynamic, target engagement and patient

---

Financial and Competing Interest Disclosure Both authors are full-time employees of Merck & Co. and own Merck stock.

K.C.N. Chang (✉) · M.J. Marton
Translational Molecular Biomarkers, Merck & Co., Rahway, NJ, USA
e-mail: ken.chang@merck.com

R. Weiner and M. Kelley (eds.), *Translating Molecular Biomarkers into Clinical Assays*, AAPS Advances in the Pharmaceutical Sciences Series 21,
DOI 10.1007/978-3-319-40793-7_15

stratification biomarkers to answer these critical questions. Genomic biomarker assays, defined here as any assays that utilize RNA- or DNA-based technologies, are crucial tools in the hands of clinical drug development teams needing to address these types of questions. In general, the clinical biomarker assay development process starts when the drug development team proposes to use a biomarker identified from preclinical experimental data (or other rationale) in a human clinical study and begins to develop a bioanalytical method (assay) to measure the biomarker. Once the assay is developed, its pre-analytical and analytical validity will be confirmed through a series of analytical validation experiments that will document its performance characteristics. A key question in the validation of any biochemical assay that will detect or quantify the biomarker is "How will the data from this assay be used in the clinical trial study?" Based on the intended use, members of the team will conduct an evaluation of the assay's accuracy, precision, limits of quantification, biological variability, stability, etc., using control samples and samples that mimic real clinical samples. As soon as the analytical validity of the assay is established, the assay is ready for use in the clinical study. If the assay is predictive or prognostic in nature, i.e., used in patient stratification or is a disease biomarker, the clinical validity, which requires clinical outcome data from the clinical trial, will be

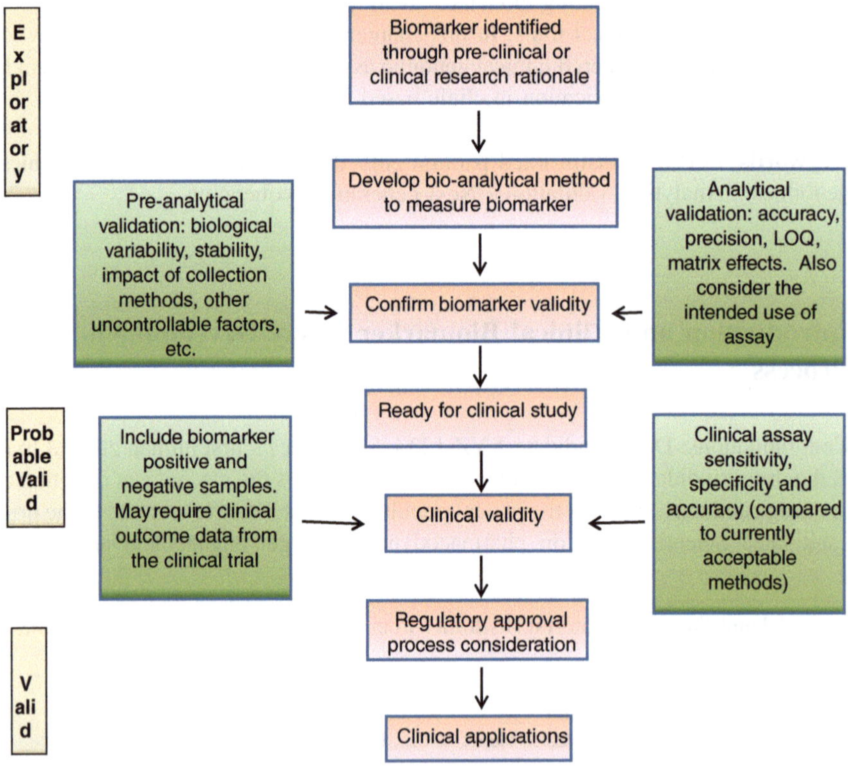

Fig. 1 Flow chart of the clinical biomarker development process

evaluated by establishing the assay's sensitivity, specificity, accuracy (compared with currently accepted methods or gold standard assays if available), and reference range for a given population (Fig. 1). The regulatory approval process is very important and will certainly impact both the selection of clinical assay platform and the validation strategy; however, a discussion of regulatory considerations is beyond the scope of this chapter and will be covered in other chapters in this e-book. Throughout this biomarker development and validation process, many factors will influence the final selection of biomarker detection platforms. In this chapter, we discuss the genomic biomarker assay technologies most commonly used in the field of clinical genomics, focusing on particular technologies with which we have extensive experience and those which we feel offer advantages over traditional genomic technologies. We conclude with a discussion of key challenges in technology platform selection, translatability across genomic technology platforms, and approaches to overcoming sample quality challenges.

Genomic Biomarker Types and Their Associated Detection Techniques

A comprehensive list of genomics assay technologies would be very lengthy. However, among these technologies only a handful of them are commonly used in clinical trials and fewer yet ultimately reach the status of a regulated clinical diagnostics product. Among the most common technologies deployed in clinical studies are quantitative PCR (qPCR), Sanger sequencing, microarray technology, and fluorescent in situ hybridization (FISH). More recently, several additional platforms, such as the Nanostring nCounter® system, mass spectrometry, and Next Generation Sequencing (also known as Massively Parallel Sequencing) technologies have started to enter the clinical testing field. Clinical genomic assays also can be placed into two broad categories: those that measure RNA/miRNA transcript abundance (including differential RNA expression signatures) or localization (such as chromogenic in situ hybridization) and those that determine DNA sequence of mutations or single nucleotide polymorphisms (SNPs), or gene copy number. However, in many cases the same technology can be used for either biomarker category. In the following several paragraphs, we describe each technology in more detail and share our experiences, recognizing that readers will have to make their own decision regarding which technology is most suitable for their biomarkers in their particular circumstances.

Quantitative PCR (QPCR)

The most frequently employed technology for genomics assays is qPCR. There are literally hundreds of commercially available qPCR assays in different forms

Table 1 Summary of frequently encountered clinical genomic technologies

Frequently encountered genomics technologies	Genomic Applications	Pros	Cons
qPCR	mRNA and miRNA gene expression, SNP typing, mutation detection	Most widely available assays, clear path for FDA clearance, considered by some as gold standard, low cost, fast TAT	Data consistency for FFPE tissue samples
Sanger sequencing	Mutation detection, SNP typing	Well established and widely available technology	Low detection sensitivity
DNA microarrays	mRNA and miRNA gene expression, SNP typing, mutation detection	Genome-wide information generation	Cost
In situ hybridization	DNA copy number, mRNA gene expression	Reliable, specific, signal localization	Limited information obtained
Nanostring technologies	mRNA and miRNA gene expression, DNA copy number	FFPE tissue sample-friendly, fast TAT, avoid amplification bias	Cost, maximum 800 genes
Mass spectrometry	Mutation detection, SNP typing	High sensitivity, less potential amplification bias than qPCR	Cost
Next generation sequencing	mRNA and miRNA gene expression, SNP typing, mutation detection, DNA copy number	"Comprehensive" view of data, most versatile, generates a lot of valuable data	Cost, difficult to validate, TAT, need for extensive data analysis, generates a lot of valuable data

designed for various purposes, and we will make no attempt to review them all. Instead, we will provide a few examples based on our experience to illustrate that qPCR is very popular for gene expression profiling assays, SNP assays, and mutation detection assays. For gene expression, the expression level of each gene could be quantified through quantitative RT-PCR such as Taqman™ or SYBR Green™ assays. The differential expression of target genes can be used as a pharmacodynamic (PD) marker [1] or for patient stratification purposes [2]. SNP assays are also commonly used as the clinical trial tests and in many cases as diagnostics tests in which allele-specific primers are used in a Taqman™-like qPCR assay [3]. This application grew out of many genome-wide association studies (GWAS) that revealed correlations between SNPs and diseases, or SNPs and side effects, since in some cases patients with different SNP subtypes (such as VKORC1, CYP2C9, etc.) showed side effects with certain drug treatments due to the SNPs' influence on the drug metabolism [4]. Recently, many new qPCR-based mutation detection assays have been approved by the FDA, including some as companion diagnostics (references here or below), i.e., tests required to determine if a given patient will likely benefit from a specific therapeutic. The most well-known mutation detection assays include Roche's cobas™ BRAF V600 Mutation Test [14] and cobas™ EGFR Mutation Test [15], Qiagen's Therascreen™ KRAS RGQ PCR Kit [16] and Therascreen™ EGFR RGQ PCR Kit [17], and very recently bioMerierux's THxID™ BRAF Test [18]. Other custom designed mutation assays that take advantage of PCR as part of their assay procedures include a single nucleotide primer extension (SNPE) assay that has been used in the recent clinical trial for patient enrollment decision-making process [5]. Quantitative PCR such as digital PCR can be used to determine DNA copy number for gene amplification detection which has implications in cancer and other diseases and which is routinely used for detection of copy number changes in CYP2D6 [6, 7]. In Table 1, we summarized the pros and cons of the frequently encountered clinical genomic technologies.

Sanger Sequencing

Sanger sequencing is another frequently encountered genomics assay for SNP detection and somatic mutation detection purposes in the clinical trial setting. Because Sanger sequencing is regarded as a gold standard for DNA sequencing only very limited assay validation efforts are required in order to be used as a clinical trial assay [8, 9].

DNA Microarrays

Microarray technologies became a very popular genomic research tool in the last 15–20 years but only more recently have they been deployed as quantitative clinical assays. One of the most well-established microarray technologies is the Affymetrix GeneChip™, which employs semiconductor manufacturing technologies [19]. In the clinical setting, this technology can be used to generate (1) an RNA expression profile using an Affymetrix whole human genome array, (2) a whole-genome SNP profile, or (3) a Cytochrome P450 profile (Roche AmpliChip P450 Genotyping test) [20], or (4) a gene sequence by re-sequencing DNA (Roche AmpliChip P53 Test, a test that detects P53 mutations) [21]. The clinical applications using microarrays include gene signature profiling for patient stratification or even prediction of treatment outcomes, exploratory whole-genome SNP profiling, or gene mutation detection for patient stratification.

In Situ Hybridization (ISH)

Fluorescent in situ hybridization (FISH) is a cytogenetic assay used to detect and potentially quantify the copy number of DNA sequences on chromosomes. In the past, it has been used in research settings to detect specific RNA targets in formalin fixed paraffin embedded (FFPE) tissue slides, but newer technologies have permitted use in clinical studies. One of such emerging technologies, called RNAscope™, is a novel RNA in situ hybridization (ISH) technology that allows signal amplification to achieve the detection sensitivity required for a clinical-grade ISH assay [10]. A full discussion and consideration of its application of ISH is covered in another chapter in this e-book.

NanoString Technologies

Although NanoString technology has been commercially available since 2008, only recently has the technology gained favor with clinical researchers. The unique feature of this technology is that it does not involve PCR amplification, instead uses hybridization-based design and high-density fluorophore as its detection methodology [22]. Because it uses very short target hybridization sequences (as short as 50 target bases available for the probes), this direct hybridization methodology is able to circumvent the challenges associated with using fragmented, and in many cases highly degraded, FFPE tissue samples, which pose difficulties for all other technologies including Sanger sequencing, qPCR and microarrays. The technology is also well suited as a clinical diagnostic platform as demonstrated by the recent FDA clearance of their assay for breast cancer typing [23].

Mass Spectrometry

The use of mass spectrometry for genomic assays in the clinic is one of the up-and-coming applications of this technology and Sequenom, a major manufacturer, plans to seek FDA clearance for its clinical diagnostics platform [24]. The Sequenom Mass Array™ is an example of such a technology, which has been used to detect DNA sequence mutations. Its detection sensitivity is generally thought to be between 5 and 10 %, which is more sensitive than Sanger sequencing, and it recently has become a useful tool to confirm mutation calls made by Next Generation Sequencing assays [11]. This technology is covered in greater detail by another chapter in this e-book.

Next-Generation Sequencing (NGS)

The most recent genomics assay technology and perhaps the technology that has the greatest potential to revolutionize the clinical genomics field is next-generation sequencing (NGS) which involves massively parallel sequencing techniques such as ion conductor and reversible-terminator chemistry, etc. [11]. This technology has quickly become a widely used research tool in preclinical research. Despite the many challenges that still need to be addressed the technology is already being used in clinical practice and in clinical trials, especially for patient stratification or diagnostic purposes. Academic labs were the first to implement the technology in a clinical laboratory and now dozens of commercial labs offer testing services that include whole exon sequencing, targeted re-sequencing and oncogene hotspot mutation detection assays. Some of the challenges include false positive calls caused by the clonal amplification (which adversely affect accuracy and reproducibility), different and inconsistent methodologies for library preparation that could lead to different types of errors, and the lack of inter-platform (such as ion conductor versus reversible terminator chemistry) data concordance and how to resolve the discrepancies [11]. On top of all these challenges, data analysis and data management issues might present the biggest obstacle that could hold or delay these transformative technologies from being widely used in the clinic.

Technology Platform Selection Requires Balancing Study Requirements

In most situations, there are likely several technologies that could potentially deliver an assay that detects a given specific RNA- or DNA-based biomarker. Therefore, determining which technology will provide the best overall performance may depend on the clinical trial objectives or other special conditions, or even tissue types, in addition to the performance characteristics of the platform itself. For example, a clinical study may have the seemingly opposing objectives of using a defined RNA gene signature to enroll patients while also requiring a global transcript assay to further develop or refine the predictive gene signature. The first objective may lend itself more to a qPCR-based assay, whereas a microarray or RNA-Seq assay may be more appropriate for the latter objective. In other words, platform selection often requires balancing conflicting requirements. Some of the technical and logistic options that need to be considered for genomics assay platform selection are discussed below.

Potential Issues When Translating from One Platform to Another

One important issue that needs to be considered when selecting a technology for developing a clinical trial assay is the translatability of the biomarker from one platform to the other. For example, most RNA gene signature biomarkers are identified through microarray studies. Yet, there are very real concerns about using microarray technology in the clinical setting, including the challenge of analytical validation, the turn-around time, and the cost. Generating global transcript profiles, but then only use a handful of the transcripts for a gene signature score calculation might be over-kill. Thus, in some cases, once the microarray-derived gene set and signature scores are defined and show potential clinical value in, e.g., patient stratification or diagnostics, the decision is generally made to migrate this set of signature genes into a more manageable or less complex technology platform, such as qPCR. Furthermore, qPCR technology has a clear path to move forward as a clinical genomics assay since the analytical validation strategy is more well-defined and accepted. However, achieving the translatability of the biomarker, as demonstrated by a good correlation between microarray and qPCR, may not be as straightforward as one would like to think, especially when FFPE tissue samples are involved. For example, the biomarker may not be directly translatable due in part to the nature of microarray design. In most cases, the first step in translating a gene set identified from microarray to the qPCR assay is to identify a set of pre-designed individual gene assays and then determine the degree of correlation between the microarray and qPCR data. However, the Affymetrix GeneChip™ uses more than 10 probe sets to define one transcript and uses a specific algorithm to calculate signal intensity, whereas the qPCR probe set design uses only one set of probes to represent a specific gene or transcript. Since the 10 plus probe sets that define a transcript typically stretch across hundreds of bases, it is common for there to be considerable variability in the signal intensity of the probe sets. This variability may cause poor correlation between microarray and qPCR data. In addition, occasionally, mismatch probes might show higher intensities than those from perfect match probes suggesting that, depending on the region of design, different probes targeting the same transcript could have different detection sensitivities. Therefore, a researcher could observe poor translatability of the biomarker if the qPCR primer/probe sets are not designed from the microarray probe sequencing covering the exact section of the transcript on the array.

Clinical FFPE Tissue Samples Require Special Attention

FFPE samples are commonly used in oncology for diagnosis and histology. When the translation and validation of a biomarker from one platform to another involves FFPE tissues, extra precautions need to be taken to ensure the extent of fragmentation from the formalin fixation will not adversely impact the translation of the biomarker. DNA extracted from FFPE tissue samples typically has a median size of around 150 nucleotides. If a primer/probe set to be used to detect a transcript covers longer than 150 target bases, the likelihood of good detection sensitivity from this primer/probe set will not be very high, especially if one considers the fact that randomly fragmented 150 bases may only have part of the target sequence even if you specifically design the Taqman™ primer/probe with target amplicon size of 60 bases. Therefore, the second step in converting a microarray-identified gene set into a qPCR assay is to make sure smaller target sequences are used in the design of the qPCR assay primer/probe set. For a Taqman™ qPCR assay, primer/probe design normally requires a target sequence of longer than 60 nucleotides to allow the primer and probe to hybridize properly plus sufficient room for DNA polymerase to initiate replication with its normal fidelity, processivity, and exonuclease activity. In reality, not all transcripts will allow effective design of such small target sequences since there are other design constraints, such as melting temperature and secondary structure conformation. In our experience for FFPE samples, we aim to keep amplicons as close to 60 nucleotides as possible and have had poorer success as amplicon size approaches 100 nucleotides.

NanoString Technology Is a Promising Solution for Clinically Challenging FFPE Tissue Samples

NanoString technology utilizes a hybridization-based approach, and employs short target sequences in its probe design. Although approximately 70 bases are typically used to accommodate both the capture probes and reporter probes with no gaps between the two probes, the technology actually only needs as few as 50 bases for these probes to anneal properly to the target transcripts. As a result, we have observed much higher correlation between the microarray data and NanoString data ($r = \sim 0.95$) than the correlation of a microarray-qPCR for a gene signature in FFPE tissue samples.

Most of signature gene sets are identified through microarray studies. However, for FFPE tissue samples, Nanostring technology may be a superior option for gene signature expression studies. Not only does it correlate well with the original microarray gene signature data, it might even perform better than microarray itself. This is because Nanostring gene signature data are more insensitive to variation in clinical sample quality (internal unpublished studies).

Avoiding Common Mistakes in Genomic Biomarker Assay Development

One challenge frequently encountered during clinical sample testing is that many samples fail to meet minimum sample quality and quantity requirements. Efforts throughout the biomarker community typically focus on acquiring high quality pre-clinical and clinical bio-samples in order to facilitate biomarker discovery and development. However, from a clinical assay development point of view, this approach may have unintended consequences, since it may lead to assays that are insufficiently robust to meet performance requirements with actual clinical samples. Without addressing the sample collection and storage issues at the clinical sites that contribute to poor sample quality, the use of high quality samples for biomarker discovery and development may make it difficult to reproduce an observed biomarker effect in a clinical setting. Therefore, even early in the biomarker discovery-development stage, preclinical samples should possess quality characteristics that mimic those of clinical samples expected from a clinical trial. This could also be done by using preclinical samples that go through controlled degradation/fragmentation process and use them for the preclinical biomarker validation.

Potential Issues for MiRNA Biomarker Developments

In the previous sections, we discussed the potential issues related to the mRNA expression when FFPE tissue samples are involved. On the other hand, the challenges for miRNA expression profiling studies have very little to do with FFPE RNA fragmentation and quality. Instead, biased pre-amplification and amplification steps in miRNA qPCR assay protocols are the concern. Based on our unpublished study, miRNA expression correlation between Nanostring and qPCR data was relatively poor. Interestingly, even using two different versions of the same platform (ABI Taqman Low Density Array miRNA version 1 vs. version 2), the correlation was lower than anticipated ($r = \sim 0.9$), suggesting the existence of biased amplification (unpublished internal studies). Hybridization is a thermodynamics-driven process, while PCR is more of a kinetics-driven process. Any small initial bias among the hundreds of primer sets will result in a very significant difference in abundance for the PCR products at the end of PCR cycles. Although some publications have shown reasonably good miRNA expression profiling correlation between qPCR and Nanostring platforms [12], a smaller subset of miRNAs were compared and no "truth" was established to prove which platform represent more accurate measurement of the true miRNA expression level. Further investigation may be needed to compare the reliability of miRNA expression profiling via direct hybridization versus PCR amplification. Another issue with miRNA biomarker discovery and development is the inconsistency of results due to extreme biological variability of miRNA expression and regulation in many sample sets. Because there is no known stably expressed reference miRNA (which could serve as a housekeeping miRNA), miRNA assays typically lack a reliable and reproducible normalization method. This represents another major issue in miRNA expression profiling data analysis. More clinical validation might be needed to prove miRNA could become one of the key clinical genomic biomarkers.

Lessons Learned Throughout Clinical Genomic Assay Development Process

Because different biofluids are collected or preserved using different methods, it may be necessary to take sample type into account when selecting the most appropriate biomarker assay platforms. Frequently encountered clinical sample types include FFPE tissue, blood, plasma, cerebrospinal fluid (CSF), sputum, tongue scrapping, nasal scrape, core needle biopsy (CNB), fine needle aspirate (FNA), hair, follicle, bronchial brushing, skin biopsy, peripheral blood mononuclear cells (PBMC) and bone marrow. The collection method, the methodology/procedure of sample stabilization, the quality and amount of DNA/RNA available, potential bacteria contamination, the stability, and variability of the stimulating agents used all need to be taken into account for the downstream assay. For analysis of cancer tissues in FFPE, a macro-dissection procedure should be included when appropriate, and of course needs to be well documented and consistent throughout the analytical validation and clinical sample testing. Many of the special considerations for clinical sample collection and preservation in general are covered in separate chapters in this e-book. We wish to emphasize one additional important point: the use of local hospital laboratories instead of a centralized testing laboratory introduces significant risk into the analysis of clinical study samples. That is, data concordance between local and central labs can be poor and can put patients and study objectives at risk [13, 14], even if FDA approved assays were used (unpublished observations). One possible contributing factor is stated above: that the inconsistent results generated in the clinic may be due to the existence of lower quality clinical samples which were not anticipated because high quality clinical mimetic samples were used in the various steps of validation. In order to minimize these unintended consequences, drug development teams should make sure that fit-for-purpose analytical validation truly is fit-for-purpose. Alternatively, one could routinely include a step of evaluating the assay performance using a sample set with wide range of sample quality to understand the potential limit of each assay, including those that would fail sample acceptance criteria.

Conclusion

Clinical assay development is quite different from preclinical assay development, since a clinical assay requires special considerations to make sure a biomarker discovered preclinically will be translatable into clinical settings. In this chapter we provided a brief overview of most frequently encountered clinical genomic assays and dissected many of them to show their practical application in clinic based on our experiences. We hope readers will find the discussion of the practical issues and lessons learned throughout the complete clinical genomics biomarker assay development process informative and helpful for avoiding problems in their clinical genomics biomarker assay development endeavor.

Future Perspective

As stated in the beginning of this chapter, a comprehensive list of genomics assay technologies would be very lengthy, however only a handful of them are commonly come across in the clinical trials or become regulated clinical diagnostics. These genomic assay technologies have proven their practical value in the clinic. The advent of next-generation sequencing has brought great excitement to the field of clinical genomics and this technology may eventually replace some types of currently popular biomarker assays. However, the scientific community will have to overcome the challenges listed earlier before that happens.

Key Terms

FFPE tissue samples
FFPE (formalin fixed paraffin embedded) tissues are one of the most frequently encountered solid tumor tissues to be used in the clinic

Gene signature
A group of genes in a type of cell or tissue with unique expression pattern characteristic that could be used to identify a medical condition or predict treatment response.

Key Learning Points

Clinical assay platform migration
Translatability of a biomarker from one platform to the other is an important issue that needs to be considered when selecting a technology for developing a clinical trial assay.

Lessons learned throughout clinical genomic assay development
The inconsistent results generated in the clinic may be due to improper validation that led to an assay that is not sufficiently robust to the variability and low quality of clinical samples.

Summary

1. Once the assay is developed, its pre-analytical and analytical validity will be confirmed through a series of analytical validation experiments that will document its performance characteristics.
2. Among the most commonly seen technologies in clinical genomics are qPCR, Sanger sequencing, microarray, fluorescent in situ hybridization (FISH) and those very recently started to enter the clinical testing field such as Nanostring technologies, mass spectrometry, and Next Generation Sequencing technologies.
3. Recently, many new qPCR-based mutation detection assays have been approved by FDA, including some as companion diagnostics, i.e., tests required to determine if a given patient will likely benefit from a specific therapeutic.
4. Nanostring technology uses a direct hybridization methodology with very short target hybridization sequences that could circumvent the challenges of using fragmented and highly degraded FFPE tissue samples.
5. We observed a higher correlation between the microarray data and Nanostring data than that of microarray and qPCR for gene signature in FFPE tissue samples.

6. One challenge frequently encountered during the clinical sample testing is that many samples fail to meet minimum sample requirements, either due to the low quality or due to insufficient quantity.
7. Special considerations are often needed for selecting biomarker assay platforms depending on the type of tissues used in the clinical trials.

References

1. Arai M, Jain S, Weaver AA et al (2010) Development and application of a biomarker assay for determining the pharmacodynamic activity of an antagonist candidate biotherapeutic antibody to IL21R in whole blood. J Transl Med 8:51
2. Kratz JR, He J, Van Den Eeden SK et al (2012) A practical molecular assay to predict survival in resected non-squamous, non-small-cell lung cancer: development and international validation studies. Lancet 379(9818):823–32
3. Lin CY, Chen JY, Lin TN et al (2011) IL28B SNP rs12979860 is a critical predictor for on-treatment and sustained virologic response in patients with hepatitis C virus genotype-1 infection. PLoS ONE 6(3):e18322
4. Chaudhry AS, Urban TJ, Lamba JK et al (2010) CYP2C9*1B promoter polymorphisms, in linkage with CYP2C19*2, affect phenytoin autoinduction of clearance and maintenance dose. Pharmacol Exp Ther 332(2):599–611
5. Chang KCN, Galuska S, Weiner R et al (2013) Development and validation of a clinical trial patient stratification assay that interrogates 27 mutation sites in MAPK pathway genes. PLoS ONE 8(8):e72239
6. Ramamoorthy A, Flockhart DA, Hosono N et al (2010) Differential quantification of CYP2D6 gene copy number by four different quantitative real-time PCR assays. Pharmacogenet Genomics 20(7):451–4
7. Hosono N, Kato M, Kiyotani K et al (2009) CYP2D6 genotyping for functional-gene dosage analysis by allele copy number detection. Clin Chem 55(8):1546–54
8. Pont-Kingdon G, Gedge F, Wooderchak-Donahue W et al (2012) Biochemical and molecular genetic resource committee of the college of american pathologists. design and analytical validation of clinical DNA sequencing assays. Arch Pathol Lab Med 136(1):41–6
9. Zoccoli MA, Chan M, Erker JC et al. (2008) Nucleic acid sequencing methods in diagnostic laboratory medicine, approved guideline. CLSI MM9-A 24(40). ISSN 0273-3099; Wang F, Flanagan J, Su N et al. (2012) RNAscope, a novel in situ RNA analysis platform for formalin-fixed, paraffin-embedded tissues. J Mol Diagn 14(1):22–29
10. Tzaphlidou M, Papaloukas C, Roukos DH (2012) Integrating NGS and third generation sequencing technologies into clinical genomic medicine. Future Med 12:341
11. Kolbert CP, Feddersen RM, Rakhshan F et al (2013) Multi-platform analysis of MicroRNA expression measurements in RNA from fresh frozen and FFPE tissues. PLoS ONE 8(1):e52517
12. André T, Blons H, Mabro M et al (2013) GERCOR. Panitumumab combined with irinotecan for patients with KRAS wild-type metastatic colorectal cancer refractory to standard chemotherapy: a GERCOR efficacy, tolerance, and translational molecular study. Ann Oncol 24(2):412–9
13. Bellon E, Ligtenberg MJ, Tejpar S et al (2011) External quality assessment for KRAS testing is needed: setup of a European program and report of the first joined regional quality assessment rounds. Oncologist 16(4):467–78

Websites

14. FDA News Release (2011) FDA approves Zelboraf and companion diagnostic test for late-stage skin cancer. News & Events. http://www.fda.gov/NewsEvents/Newsroom/PressAnnouncements/ucm268241
15. FDA News Release (2013) FDA approves first companion diagnostic to detect gene mutation associated with a type of lung cancer. News & Events. http://www.fda.gov/NewsEvents/Newsroom/PressAnnouncements/ucm352230
16. FDA Approves Qiagen's KRAS Companion Diagnostic and New Use for Erbitux. http://www.genengnews.com/gen-news-highlights/fda-approves-qiagen-s-kras-companion-diagnostic-and-new-use-for-erbitux/81247009/
17. QIAGEN Receives FDA Approval for therascreen® EGFR RGQ PCR Kit as a Companion Diagnostic for Lung Cancer Patients http://www.prnewswire.com/news-releases/qiagen-receives-fda-approval-for-therascreen-egfr-rgq-pcr-kit-as-a-companion-diagnostic-for-lung-cancer-patients-215235241.html
18. FDA News Release (2013) FDA approves two drugs, companion diagnostic test for advanced skin cancer. News & Events. http://www.fda.gov/NewsEvents/Newsroom/PressAnnouncements/ucm354199
19. How Affymetrix GeneChip DNA Microarrays Work http://public.tgen.org/tgen.org/downloads/autism/Genotypingessentials.pdf
20. http://www.fda.gov/MedicalDevices/ProductsandMedicalProcedures/DeviceApprovalsandClearances/Recently-ApprovedDevices/ucm078879.htm (Roche AmpliChip)
21. Merck & Co. Teams with Roche to develop companion diagnostics for cancer therapy. http://www.genengnews.com/gen-news-highlights/merck-co-teams-with-roche-to-develop-companion-diagnostics-for-cancer-therapy/81245257/
22. NanoString Technologies, http://www.nanostring.com/
23. NanoString Wins FDA Approval of Breast Cancer Diagnostic Test http://www.xconomy.com/seattle/2013/09/09/nanostring-wins-fda-approval-breast-cancer-diagnostic-test/
24. Sequenom Technologies, http://www.sequenom.com/

Author Biographies

Ken CN Chang is Senior Principal Scientist in Clinical Biomarker Development Laboratory, Merck & Co. He is the scientific lead for many clinical genomics assay development and validation projects. Prior to Merck, he worked at Wyeth Research and Pfizer Inc. for more than 17 years as principal investigator/drug discovery team leader in various disease areas, and as genomics core lab manager. His research interests and expertise include DNA structure conformation, DNA replication and repair mechanisms, chemical carcinogenesis and mutagenesis, anticancer/antiviral drug discovery, target identification/target validation using gene expression profiling technologies, as well as clinical biomarker assay development and validation.

Matthew Marton is Director of Clinical Genomics in Merck & Co.'s Clinical Biomarker Development Laboratory, where his team has deployed predictive biomarkers as patient enrichment assays on multiple technology platforms in several dozen Merck clinical studies. A current focus of his work is the CoDx strategy for novel genomic assays as molecular diagnostics, and he is a contributor to several CLSI documents involving molecular diagnostic testing in oncology. Prior work at Merck focused on technology/method development and biomarker discovery, with an emphasis on quality control of molecular profiling assays.

Implementation of Immunohistochemistry Assays for Clinical Trial Sample Analyses

Marisa Dolled-Filhart, Usha Singh, Dianna Wu and K. Emancipator

Abstract Immunohistochemistry (IHC) is an essential tool for visualization and assessment of protein expression in sections of tissue in the context of sample morphology. IHC is utilized in preclinical applications to determine target engagement and pharmacodynamics, as well as proof of concept studies. However, transitioning an assay from the preclinical stage to implementation for human clinical trial sample analyses can be challenging for a number of reasons. For example, correlation of preclinical results on surrogate tissues may not be indicative of the final results in humans and assay conditions and reagents may need to be changed for optimal use on human formalin-fixed paraffin-embedded (FFPE) tissue. In clinical trials, IHC can be deployed to address different types of biomarker needs, such as for prospective patient enrollment where the assay validated to stringent clinical laboratory standards, or as a research assay used solely to generate observational data. Using a chromogen-based oncology IHC biomarker analyses as an example, this chapter will cover a range of considerations for implementation of IHC assays in clinical trials, from sample collection considerations through assay validation and implementation.

Sample Collection and Impact of Variable Pre-analytic Conditions

It is well established that pre-analytic conditions can impact the resulting biomarker analyses when performing IHC. Potential sources of pre-analytic variables can be grouped into several main categories, as reviewed by Engel and Moore [1] including: (a) pre-fixation, (b) fixation, (c) post-fixation, (d) tissue processing into a FFPE block, (e) FFPE block sectioning, and (f) storage of blocks and slides. However, the variables with seemingly the greatest impact on pre-analytic vari-

M. Dolled-Filhart (✉) · U. Singh · D. Wu · K. Emancipator
Translational Biomakers, Merck & Co., Rahway, NJ, USA
e-mail: marisa.dolled-filhart@merck.com

R. Weiner and M. Kelley (eds.), *Translating Molecular Biomarkers into Clinical Assays*, AAPS Advances in the Pharmaceutical Sciences Series 21,
DOI 10.1007/978-3-319-40793-7_16

ability are time from removal of sample to placement in fixative, type of fixative, and as length of time in fixative due to challenges in controlling this in general practice [2] which may have greater impact on phospho-epitopes [3–5].

Sample Collection

In clinical trials, pre-analytic variables extend even further back to determining what type of sample to collect (archival tissue or newly obtained biopsies), and to the specific details of shipping and handling methods from clinical site to the laboratory that will handle/process the samples either into FFPE blocks or prepare sections from blocks for IHC testing. Collection of archival tissue versus newly obtained biopsies has different advantages and disadvantages. It is important to be aware of the differences in sample size potentially available from large tissue resections versus small core biopsies.

Some sites participating in clinical trials may prefer to prepare their own FFPE blocks or sections for use in IHC analyses. As mentioned by Hewitt et al. [2], it may commonly be stated that "tissue has been fixed and paraffin-embedded per standard protocol." That statement can refer to a variety of different protocols within or between institutions given the range of different containers, times, reagents, methods, equipment, temperature, storage, and handling factors that may be used during the course of a clinical trial. As each institution may have their own SOPs and processes for those components, it is important to clearly define and harmonize sample collection and preparation with the pathology contract research organization (CRO) so that all samples are handled as similarly as possible. As some sites (or specific countries) may have policies about not releasing paraffin blocks for clinical trials, attention to detail regarding sectioned slide preparation, shipping, storage, and time from slide sectioning to IHC analysis are very important details.

Sample Fixation

In particular, time of fixation of tissue is a key pre-analytic variable covered in the American Society of Clinical Oncology (ASCO)/College of American Pathologists (CAP) guidelines for IHC analysis of hormone receptors [6] and HER2 [7]; recommendation for breast cancer sample fixation time in 10 % neutral buffered formalin (NBF) is 6–72 h for estrogen receptor (ER) and progesterone receptor (PR) analysis, and 6–48 h for HER2 testing (due to limited studies beyond 48 h for HER2). Studies such as Goldstein et al. [8] provide insight into minimum formalin have evaluated extension of time in fixative to 72 h or 96 h without deleterious effect for specific biomarkers [9, 10]. This has relevance to the challenges of timing from sample collection through shipping and receipt at CRO for newly obtained biopsies in global clinical trials.

Time to fixation (from sample removal to placement in fixative) has also been the focus of many studies, ASCO/CAP guidelines recommend cold ischemic time of less than 1 h [6]. Other studies have addressed this issue using tissue microarrays containing samples with a range of cold ischemic times and/or matched biopsies and resections as a surrogate to short (biopsy) and longer (resections) ischemic times, and in particular there are concerns about phospho-epitopes [5, 8, 11], further summarized by Siddiqui et al. [4]. These and other studies are also complicated by the impact of tissue heterogeneity on results/interpretation. When possible, recording the time of tissue collection and time sample has been placed in fixative can help determine ischemic time, however, this adds extra burden to sites in sample collection for clinical trials. Typically, greater concern is placed on time to fixation and/or incomplete fixation due to rapid turnaround times requested in clinical settings, as compared to lesser concerns about over fixation [8].

Sample Stability

Given that trials may extend over years, time from sectioning to analysis is of particular importance given that (unstained) sectioned slide stability, also known as cut slide stablity, may only be established for days, weeks, or months and dependent on the tumor tissue and/or analyte of interest [12, 13]. The mechanism for loss of antigenicity in tissue sections from formalin-fixed paraffin embedded samples is not fully understood, but some studies have explored both methods of preventing loss by alternative slide storage methods, such as paraffin-dipping coating [14], as well as exploring the role of humidity and endogenous water retention in sectioned slides along with vacuum/desiccant storage [15]. Other methods that have been explored are storage at different temperatures, storage in nitrogen chambers and other additives to paraffin coating; loss of antigenicity for IHC within the block has not been seen as critical of a problem (reviewed in [15]). The ASCO/CAP HER2 and ER/PR guidelines state that storage of sectioned slides for more than 6 weeks is not recommended [6, 7]. Determination of sectioned slide stability over weeks and/or months may need to be determined based on specific clinical trial logistics and timing.

Guidelines

While guidelines have been developed specifically for clinical sample analysis of IHC for commonly used breast cancer IHC tests for hormone receptors [6] and HER2 [7] as described above, guidelines related to clinical trial sample collection are just starting to be defined and published. Some steps taken to harmonize these efforts are joint recommendations from the Breast International Group (BIG) and the National Cancer Institute (NCI)-sponsored North American Breast Cancer Cooperative Groups (NABCG) [12] for specimen collection and handling recommendations. Major recommendations include focus on submission of FFPE blocks, use of the same fixative across trials (10 % NBF pH 7), and handling and shipping details for FFPE tissue [12]. Other proposals for uniform biospecimen collection for breast cancer trials from BIG and NABCG have included recommendations for number of time points of biopsy specimen collection as well as minimum tissue quantities and sizes [16]. This proposal included breast cancer specific recommendations such as (a) collection of 0.8–1.2 mm biopsies with 14–16 gauge needles and (b) 4 core biopsy samples are feasible to collect per time point based on previously performed trials [16].

IHC Methodology

Staining Methods

Today's methods of IHC include both fluorescence and chromogenic-based staining; this chapter focuses on chromogen-based methods due to their higher prevalence and use in clinical trials as well as hospital labs (in general) for performing IHC; fluorescence-based IHC may also be referred to as immunofluorecence (IF) methodology. As reviewed by Matos et al. [17], there are a range of detection methods available for immunohistochemistry such as the peroxidase-anti-peroxidase method and the alkaline phosphatase-anti-alkaline phosphatase method. The advent of antigen retrieval methods, secondary antibody detection methods (i.e., avidin-biotin complex, labeled streptavidin-biotin complex) and the use of chromogens [i.e., diaminobenzidine (DAB)] have added to the popularity and increased prevalence of the use of IHC, as have the increased use of autostainers, development of dual-chromogen protocols, and the ability to develop more sensitive assays using reagents such as haptens or polymers.

IHC Scoring Methods

The method of scoring used for IHC assay depends on a number of factors including whether there are existing guidelines, the subcellular localization and distribution of staining of the biomarker. Definition of what is considered "positive" by IHC will depend on method of scoring and definition of "cut-off" for positivity in fit-for-purpose validations, as well as incorporation of results from positive and negative controls (intrinsic and extrinsic, as appropriate) [17].

Common challenges can include determination of which subcellular localizations within the tumor are of biologic importance (e.g., cytoplasm, membrane, nucleus; is partial or only complete membrane staining to be considered positive), which cell types/regions are important (e.g., tumor, stroma, blood vessels, inflammatory cells, etc.), whether to incorporate information regarding the extent of tissue and/or biomarker heterogeneity (i.e., determine whether average intensity or highest intensity areas will be captured in scoring of samples). Certainly, preclinical studies can inform on the relevance of subcellular localization, however, different results may be seen in cell lines and xenograft models as compared to human tumor samples.

Examples of commonly used pathologist-based scoring methods are summarized in Table 1; Garcia Rojo et al. [18] have reviewed the many methods of digital imaging including computer-assisted and/or computer-based IHC image analysis. Digital imaging can be useful not just for archiving of H&E and stained slides, but also for coordinating additional/collaborative review of samples.

Fit-for-Purpose IHC Validation

There are many different degrees of validation, hence discussion of "fit for purpose" validation approaches are described below for IHC-based considering study objectives and intended use of the data. This chapter does not cover companion diagnostic regulatory requirements, but instead, focuses on typical use of IHC analyses for retrospective exploratory analyses or development of a prospective patient enrollment assay. Enrollment assays require the most stringent validation because they affect patient treatment decisions. Therefore, enrollment assays are validated to clinical laboratory standards and performed in accredited laboratories. In the US, these standards and accreditations are governed by the Clinical Laboratory Improvement Amendments of 1988 (can be found at www.fda.gov), commonly known as "CLIA." As the use of enrollment assays in clinical trials increase, other regulatory agencies (e.g., FDA, EMA) as well as institutional review boards and ethics committees are becoming more interested in the level validation, and may impose even more stringent standards. The specifics of regulatory guidelines or requirements around the use of IHC assays for purposes of enrollment are not addressed in this chapter; the focus is on technical aspects and considerations of IHC assay validation.

There are many different considerations regarding selection of a reagent antibody for the development of an IHC assay including different levels of confirmation as to whether the antibody is actually detecting the antigen of interest (reviewed by Bordeaux et al. [23]). Depending on the antigen of interest, the use of different antibodies to the same antigen can sometimes provide very different results as in the case of discordance between HER3 antibodies by IHC when compared to the high correlations between different estrogen receptor (ER) antibodies by IHC [24]. Lot-to-lot variations of a single reagent antibody are also an issue [25]. An IHC

Table 1 Commonly used pathologist-based IHC scoring methods

Scoring method	Scale	Description
Binary	Positive or negative	Varies depending on application; may be based on a threshold percentage of staining cells, a threshold intensity of staining, recognition of a pattern of staining, or some combination of all of these
Traditional	0–3+	Similar to binary, but three different thresholds are defined (for example, see HercepTest™ Interpretation Guide at www.dako.com or Wolff et al. [19])
Intensity	0–3+ 0 = null, 1+ = weak, 2+ = moderate, 3 + = strong (sometimes noted as 0, +, ++, or +++)	Based on intensity of staining; often defined in terms of a fixed percentage of cells of interest, e.g., 3+ means that at least *x*% of cells stain strongly, etc.
Proportion score	0–100 %	Percentage of cells of interest positive for IHC staining
"2+ or 3+" proportion score	0–100 %	Percentage of cells of interest with moderate (2+) or strong (3+) staining
"3+" proportion score	0–100 %	Percentage of cells of interest with strong (3+) staining
H-Score	0–300	H-score = Percentage * Intensity. Percentage from 0 to 100 %. Intensity o to 3: 0 (null), 1+ = weak but detectable above control, 2 = distinct, 3+ = strong (some articles include 4+ as minimal light transmission through stained nucleus [20] but 0–300 scale is more typically used)
Allred score [21]	0, 2–8	Total score = Proportion Score + Intensity Score. Proportion score = 0 (null), 1 (<1 %), 2 ($\geq$1 % to <10 %), 3 ($\geq$10 % to <33 %), 4 ($\geq$33 % to <66 %), 5 ($\geq$66 %) Intensity score = 0–3 scale
Quick score [22]	0–7	Quick score = intensity + proportion. intensity = 0–3; proportion of positive nuclei: 0 (none), 1 ($\sim$1–25 %), 2 ($\sim$26–50 %), 3 ($\sim$51–75 %), 4 ($\sim$76–100 %)

assay must be developed for the intended tissue samples by optimizing "signal to noise" through a variety of factors such as antibody concentration, antigen retrieval, and incubation times. Focusing on the developed assay that is appropriately specific to the antigen and appropriate for use for IHC, there are several areas in which validation can be performed to provide confidence in performance of the assay over time. Obviously, use and extent of evaluation of each component will depend on how the biomarker will be used—ranging from preclinical or exploratory studies, retrospective studies, or prospective studies [26]. Typical components of such validations are (a) Specificity, (b) Sensitivity, (c) Controls, (d) Precision, (e) Sectioned slide stability, and (f) Pathologist concordance. These studies usually utilize anonymous FFPE tumor bank samples without associated clinical outcome information.

Specificity and Sensitivity

Specificity is typically performed on a panel of normal tissues in order to establish the reactivity of the antibody across the wide range of normal human organs. This can often be helpful in identification of positive and negative extrinsic tissue controls to use for future staining runs. Sensitivity is typically performed on a panel of samples of the tumor of interest or across a range of indications. The number and type of samples depends on the level of validation, e.g., whether or not the assay is used for prospective patient enrollment. Sensitivity analysis allows for confirmation that an appropriate dynamic range of expression and appropriate staining localization is seen for the tumor type(s) of interest prior to proceeding to precision and sectioned slide stability studies. Depending on the assay, it may also be useful to have 3rd party confirmation/concordance of results for a given IHC marker.

Controls

It is critical to run positive and negative controls in parallel with every staining run when performing IHC as they allow for confirmation of appropriate assay performance and sample conditions. External/extrinsic controls refer to controls that are separate tissue samples than the testing clinical samples of interest. Internal/intrinsic controls refer to tissue elements within the testing samples. Unfortunately, appropriate intrinsic positive controls do not exist for all biomarkers; some studies have demonstrated use of "integrated" positive controls into the testing tissue sample in cases in which intrinsic controls are lacking. This has some benefits but also drawbacks as this require alteration of the original patient sample block [27]. While in a discovery setting, use of xenografts and/or cell lines with ranges of biomarker expression may be useful. It is important to evaluate human tissue controls in the context of clinical trial sample analysis. Processing of controls also needs to be considered when developing IHC tests that may require other types of processing (such as decalcification of bone marrow specimens).

As mentioned, the use of appropriate positive and negative controls (both tissue controls and reagent controls) is critical for each staining run for clinical trial studies. Commonly used negative reagent controls are diluent or buffer alone or a matched immunoglobulin control. Negative tissue controls lack expression of the biomarker of interest while positive tissue controls are known to express the biomarker of interest. A more in depth summary of the different types of controls and their uses are provided by Torlakovic et al. [27].

Precision, Sectioned Slide Stability, and Pathologist Concordance

Precision allows for assessment of assay reproducibility that addresses how the assay performs across multiple parameters such as assessment of intra- and inter-assay variability, and may include use of multiple operators and instruments. A typical configuration would be to choose samples for precision representing negative, low, moderate, and high expression (depending on scoring system) of the biomarker of interest based on sensitivity screening. Three serial sections of each sample would be stained in three separate staining runs. This would allow for within-run and between-run variability to be assessed, as well as any assessments between operators and instruments.

A similar configuration of samples with a range of expression of the biomarker would be selected for use in sectioned slide stability studies. These types of stability studies should be used to reflect what may be seen in collection of samples in clinical trials, such as comparing IHC results between freshly cut sections versus sections stored for various lengths of time at room temperature [13] and/or in other storage conditions (e.g. refrigeration). These results will provide further details to put into the procedure manual regarding time from slide sectioning to sending for IHC analysis, as well as allow for optimal determination of batching requirements based on sample receipt. Given variation in interpretation between pathologists, it is important to establish scoring guidelines and to ensure concordance if there will be multiple pathologists involved in scoring IHC from clinical trial samples.

Conclusions

When deploying an assay for clinical trial sample analysis, it is crucial to spend time refining sample collection and pre-analytic conditions via specific detailed instructions in a procedure manual. If working with a central lab to supply materials to sites, be very specific in which materials are required and checking through a sample kit to ensure directions and materials are compatible with procedure manual instructions. When working with a contract research organization (CRO) to perform IHC for a clinical trial, it is important to understand not just the laboratory turn-around time (TAT) but the time from sample receipt to data reporting.

TAT for real-time IHC analysis in clinical trials can vary typically from 2 to 7 business days, not including shipment time from the site to the testing laboratory. The TAT varies by laboratory, and depends on a number of factors such as IHC assay protocol length, complexity in H&E review and IHC scoring, as well as whether pathologists are on staff or are consultants. TAT for retrospective studies are based on batch size and frequency of testing. Additional time will be required if biopsies are arriving in NBF and need to be processed into a FFPE block prior to beginning testing. It is also important to determine up front any sample accessioning and tracking requirements, data transfer specifications, and database/data management requirements.

By taking into account these elements, as well as the scientific and technical components described above, pre-analytic and analytic variability can be decreased to provide greater reliability in IHC results for clinical trial sample analyses.

Key term 1 Immunohistochemistry (IHC)—detection of antigens in tissue sections based on antibody binding with protein detection via chromogen or fluorescence; this technique is widely used for sample diagnoses and biomarker analyses and is visualized via microscopy.

Key term 2 Formalin-fixed paraffin-embedded (FFPE) tissue—a commonly used method for fixing and preserving tissue that has been removed by biopsy or resection.

Key term 3 IHC scoring—the method by which the pathologist scores the results of IHC staining, which is based on one or more factors such as the cellular localization, intensity and/or percentage of staining seen in the tissue.

Summary

- IHC is an important tool for protein expression visualization in tissue samples in clinical trial sample analysis.
- Pre-analytic conditions such as sample fixation and stability, as well as sample collection instructions and materials, need to be carefully considered in clinical trial sample collection and IHC analysis.
- There are several different types of pathologist-based IHC scoring methods; selection of which method is most appropriate is based on the specific tissue type and biomarker IHC staining patterns.
- The level of validation of IHC assays is based on clinical trial study objectives and intended use of the data.

References

1. Engel KB, Moore HM (2011) Effects of preanalytical variables on the detection of proteins by immunohistochemistry in formalin-fixed, paraffin-embedded tissue. Arch Pathol Lab Med 135:537–543
2. Hewitt SM, Badve SS, True LD (2012) Impact of preanalytic factors on the design and application of integral biomarkers for directing patient therapy. Clin Cancer Res 18:1524–1530

3. Pinhel IF, Macneill FA, Hills MJ, Salter J, Detre S, A'hern R, Nerurkar A, Osin P, Smith IE, Dowsett M (2010) Extreme loss of immunoreactive p-Akt and p-Erk1/2 during routine fixation of primary breast cancer. Breast Cancer Res 12:R76
4. Siddiqui S, Rimm DL (2010) Pre-analytic variables and phospho-specific antibodies: the achilles heel of immunohistochemistry. Breast Cancer Res 12:113
5. Holzer TR, Fulford AD, Arkins AM, Grondin JM, Mundy CW, Nasir A, Schade AE (2011) Ischemic time impacts biological integrity of phospho-proteins in PI3 K/Akt, Erk/MAPK, and p38 MAPK signaling networks. Anticancer Res 31:2073–2081
6. Hammond ME, Hayes DF, Dowsett M, Allred DC, Hagerty KL, Badve S, Fitzgibbons PL, Francis G, Goldstein NS, Hayes M, Hicks DG, Lester S, Love R, Mangu PB, McShane L, Miller K, Osborne CK, Paik S, Perlmutter J, Rhodes A, Sasano H, Schwartz JN, Sweep FC, Taube S, Torlakovic EE, Valenstein P, Viale G, Visscher D, Wheeler T, Williams RB, Wittliff JL, Wolff AC (2010) American society of clinical oncology/college of American pathologists guideline recommendations for immunohistochemical testing of estrogen and progesterone receptors in breast cancer (unabridged version). Arch Pathol Lab Med 134:e48–e72
7. Wolff AC, Hammond ME, Hicks DG, Dowsett M, McShane LM, Allison KH, Allred DC, Bartlett JM, Bilous M, Fitzgibbons P, Hanna W, Jenkins RB, Mangu PB, Paik S, Perez EA, Press MF, Spears PA, Vance GH, Viale G, Hayes DF (2013) Recommendations for human epidermal growth factor receptor 2 testing in breast cancer: American society of clinical oncology/college of American pathologists clinical practice guideline update. Arch Pathol Lab Med
8. Goldstein NS, Ferkowicz M, Odish E, Mani A, Hastah F (2003) Minimum formalin fixation time for consistent estrogen receptor immunohistochemical staining of invasive breast carcinoma. Am J Clin Pathol 120:86–92
9. Yildiz-Aktas IZ, Dabbs DJ, Cooper KL, Chivukula M, McManus K, Bhargava R (2012) The effect of 96-hour formalin fixation on the immunohistochemical evaluation of estrogen receptor, progesterone receptor, and HER2 expression in invasive breast carcinoma. Am J Clin Pathol 137:691–698
10. Tong LC, Nelson N, Tsourigiannis J, Mulligan AM (2011) The effect of prolonged fixation on the immunohistochemical evaluation of estrogen receptor, progesterone receptor, and HER2 expression in invasive breast cancer: a prospective study. Am J Surg Pathol 35:545–552
11. Neumeister VM, Anagnostou V, Siddiqui S, England AM, Zarrella ER, Vassilakopoulou M, Parisi F, Kluger Y, Hicks DG, Rimm DL (2012) Quantitative assessment of effect of preanalytic cold ischemic time on protein expression in breast cancer tissues. J Natl Cancer Inst 104:1815–1824
12. Leyland-Jones BR, Ambrosone CB, Bartlett J, Ellis MJ, Enos RA, Raji A, Pins MR, Zujewski JA, Hewitt SM, Forbes JF, Abramovitz M, Braga S, Cardoso F, Harbeck N, Denkert C, Jewell SD (2008) Recommendations for collection and handling of specimens from group breast cancer clinical trials. J Clin Oncol 26:5638–5644
13. Divito KA, Charette LA, Rimm DL, Camp RL (2004) Long-term preservation of antigenicity on tissue microarrays. Lab Invest 84:1071–1078
14. DiVito KA, Charette LA, Rimm DL, Camp RL (2004) Long-term preservation of antigenicity on tissue microarrays. Lab Invest 84:1071–1078
15. Xie R, Chung JY, Ylaya K, Williams RL, Guerrero N, Nakatsuka N, Badie C, Hewitt SM (2011) Factors influencing the degradation of archival formalin-fixed paraffin-embedded tissue sections. J Histochem Cytochem 59:356–365
16. Loi S, Symmans WF, Bartlett JM, Fumagalli D, Van't Veer L, Forbes JF, Bedard P, Denkert C, Zujewski J, Viale G, Pusztai L, Esserman LJ, Leyland-Jones BR (2011) Proposals for uniform collection of biospecimens from neoadjuvant breast cancer clinical trials: timing and specimen types. Lancet Oncol 12:1162–1168
17. Matos LL, Trufelli DC, de Matos MG, Silva Pinhal MA (2010) Immunohistochemistry as an important tool in biomarkers detection and clinical practice. Biomark Insights 5:9–20

18. Rojo MG, Bueno G, Slodkowska J (2009) Review of imaging solutions for integrated quantitative immunohistochemistry in the Pathology daily practice. Folia Histochem Cytobiol 47:349–354
19. Wolff AC, Hammond ME, Hicks DG, Dowsett M, McShane LM, Allison KH, Allred DC, Bartlett JM, Bilous M, Fitzgibbons P, Hanna W, Jenkins RB, Mangu PB, Paik S, Perez EA, Press MF, Spears PA, Vance GH, Viale G, Hayes DF (2014) Recommendations for human epidermal growth factor receptor 2 testing in breast cancer: american society of clinical oncology/college of american pathologists clinical practice guideline update. Arch Pathol Lab Med 138:241–256
20. McCarty KS Jr, Szabo E, Flowers JL, Cox EB, Leight GS, Miller L, Konrath J, Soper JT, Budwit DA, Creasman WT (1986) Use of a monoclonal anti-estrogen receptor antibody in the immunohistochemical evaluation of human tumors. Cancer Res 46:4244s–4248s
21. Allred DC, Harvey JM, Berardo M, Clark GM (1998) Prognostic and predictive factors in breast cancer by immunohistochemical analysis. Mod Pathol 11:155–168
22. Rhodes A, Jasani B, Barnes DM, Bobrow LG, Miller KD (2000) Reliability of immunohistochemical demonstration of oestrogen receptors in routine practice: interlaboratory variance in the sensitivity of detection and evaluation of scoring systems. J Clin Pathol 53:125–130
23. Bordeaux JM, Cheng H, Welsh AW, Haffty BG, Lannin DR, Wu X, Su N, Ma XJ, Luo Y, Rimm DL (2012) Quantitative in situ measurement of estrogen receptor mRNA predicts response to tamoxifen. PLoS ONE 7:e36559
24. Anagnostou VK, Welsh AW, Giltnane JM, Siddiqui S, Liceaga C, Gustavson M, Syrigos KN, Reiter JL, Rimm DL (2010) Analytic variability in immunohistochemistry biomarker studies. Cancer Epidemiol Biomark Prev 19:982–991
25. Pozner-Moulis S, Cregger M, Camp RL, Rimm DL (2007) Antibody validation by quantitative analysis of protein expression using expression of Met in breast cancer as a model. Lab Invest 87:251–260
26. Marchio C, Dowsett M, Reis-Filho JS (2011) Revisiting the technical validation of tumour biomarker assays: how to open a Pandora's box. BMC Med 9:41
27. Torlakovic EE (2012) Quality control by tissue microarray in immunohistochemistry. J Clin Pathol 65:961

Managing Biomarker Outsourcing: CRO Evaluation, Streamline Outsource Process, and Quality Management

Jeffrey A. Tsou

Abstract For the past decade the pharmaceutical industry has faced numerous challenges, such as the rising cost of R&D, a thin pipeline, the patent cliff, increased regulatory restrictions, and inefficient drug development processes. The need to increase efficiency and improve drug development processes necessitates companies to outsource segments of the drug development process to CROs. The movement of outsourcing certain core competencies transformed the outsourcing strategy from a tactical outsourcing model to strategic partnered relationship. Historically, companies followed the tactical outsourcing model in order to fill the gaps and provide more capacity on an individual project on a case-by-case need. The new strategic partnered relationship model outsources facets of the drug development process based on collaboration, open knowledge sharing, and relationship development. This chapter will focus on the key aspects of success in outsourcing clinical biomarker assays to specialty CRO laboratories. There are three key steps in successful outsourcing of clinical biomarker assays; assessment and selection of CRO, application of the strategic partnered relationship outsourcing model, and trust and execution. Completion of successful outsourcing will establish a relationship that can be fruitful and rewarding for future projects and help support an innovative drug development process.

Keywords Clinical biomarkers · Strategic outsourcing · CRO · Drug development

Background

Over the past decade, the pharmaceutical industry has been challenged to contain the increased costs of research and development (R&D) efforts for drug development. The costs for developing new drugs have been climbing from an estimated $50 billion in 2008 while productivity of the drug development process has been

J.A. Tsou (✉)
Department of Reproductive and Genetic Health, Illumina, San Diego, CA, USA
e-mail: tsouwhat@gmail.com

R. Weiner and M. Kelley (eds.), *Translating Molecular Biomarkers into Clinical Assays*, AAPS Advances in the Pharmaceutical Sciences Series 21,
DOI 10.1007/978-3-319-40793-7_17

declining [1]. In addition to the decline in productivity, the numbers of total drug approvals (new molecular and biological entities) have dropped significantly in the past 20 years due to regulatory reforms and increased safety concerns [2]. To compound this issue, the patent cliff, the wave of drug patent expirations by major pharmaceutical companies, started around 2010, and further jeopardizes the sustainability of major pharmaceutical companies. Several blockbuster drugs such as Enbrel, Singulair, and Lipitor, have recently gone off-patent while others, such as Celebrex, Crestor, and Zetia will go off-patent within the next 2 years. The estimated loss of these patents are estimated to result in an estimated $250 billion sales risk within the next 5 years [3]. The return on investment is low based on the rising costs, regulatory restrictions, patent cliff, and shrinking pipeline. All of these elements are driving the pharmaceutical industry to develop new drug development strategies.

Within the last decade, many companies strived to fill their pipeline through mergers and acquisitions as part of the new drug development strategy. Pharmaceutical companies that implemented this strategy include Pfizer and Wyeth, Merck and Schering Plough, J&J and Synthes, Roche and Genentech, and Genzyme and Sanofi-Aventis. However, the results from these mergers have caused decrease in productivity, company downsizing, site closures, layoffs, and elimination of certain therapeutic areas of research. These changes in structure and strategy have also forced companies to operate with a smaller budget, seek ways to cut costs, and reduce the time for drug development. Pharmaceutical industry started to reassess the R&D strategy to create a more efficient drug development process since there are only so many opportunities through mergers and acquisitions. To address the inefficiencies described earlier, many companies looked toward Clinical Research Organizations (CRO) to outsource segments of the drug development process. Recent studies have demonstrated that projects outsourced to CROs inclined to be larger, closer to project timelines, and are associated with faster development times. As a result, more projects that were outsourced increased the CRO industry as drug sponsor by 15 % and increased the headcount at major CROs by 6 % annually [4].

As described in a recent review, outsourcing was historically done on a tactical model, a case-by-case strategy to fill the gaps in order to provide more capacity and gain additional access to research experience [5]. Over a decade ago, it was unthinkable to outsource core competencies, also known as the functions, expertise, or knowledge that provides the skills to develop competitive drugs. Ideally, it is believed that companies should not outsource their core competencies with the fear of losing the proprietary intellectual property (IP) or leading edge losing its competitiveness. Recently, there has been new outsourcing strategies performed in practice such as preferred provider relationship, functional service provider relationship, business process outsourcing, and full development outsourcing (Table 1). Initially, outsourcing in the discovery space has been slow to catch on due to the limitations of four factors: the availability of capable vendors; discovery phases were considered core competencies; industry is not up to date on discovery technologies; and outsourcing is a commitment with risky management decisions [6]. However, the lack of efficiency in the drug development process drove companies

Table 1 Pharmaceutical outsourcing models

Model	Strategy	Pros	Cons	Industry use
Tactical	Outsource selected minimal activities not related to core competencies	• Free up internal resources • Decreased cost • Speed of execution	• Limited quality on sourced projects • Increased time to choose CRO since no established relationship	Historically to fill gaps and increase capacity
Strategic partnered relationship	Partnering with a CRO selected by four C's to outsource certain core competencies on numerous projects	• Open collaboration fosters innovation and trust • Increase process cycle time • Decrease costs to make internal resources available for other R&D efforts	• Relationship takes significant amount of time to develop • Initial assessment of CROs needed and Limited to selecting from a pool of pre-selected CROs • Outsourcing core competencies may risk loss of IP or market edge	Clinical biomarker assays
Functional service provider relationship	Outsource complete services	• Partner expertise can be leveraged to reduce process cycle time and costs • Established partner allows improved quality and increased efficiencies with repeated use • Retraining and set up is not required with future projects • Can eliminate redundant internal departments	• Sponsor resource still required to oversee the process and to manage the interaction of multiple service streams • Limited to established repetitive process with defined expertise	Data management, clinical monitoring, investigator payments
Business process	Outsourcing the operations and responsibilities of a specific business function	• Internal redundant process may be eliminated • Reduce costs for noncore processes • Leverage expertise of CRO for increased efficiency	• Limited to be used for services with defined process	IT

(continued)

Table 1 (continued)

Model	Strategy	Pros	Cons	Industry use
Full development	CRO becomes extension of the sponsor in the execution of the whole molecule or specific parts of the clinical development plan	• Sponsor has access to the CROs cross industry therapeutic operational experience • Fixed price contract with shared risk and reward • Strong governance and management support • Opportunity to improve productivity and cost effectiveness • Therapeutic aligned facilitating logistics to support studies globally • Provides for capacity, flexibility, and responsiveness to a growing portfolio	• Time and effort to set up relationship • Pre-selection of a limited number of CROs • Established relationship prevents changing vendors mid process • Limited to pulling programs back internally if projects are unsuccessful	Clinical trials

Comparison of the current pharmaceutical outsourcing models, intended strategies, and processes that can be outsourced to a CRO. Adapted and expanded from [5]

to outsource core competencies and success has been reported in the outsourcing of lead optimization [7]. Specifically, success in outsourcing complete segments of the drug discovery process, such as screening, chemical intermediates, and safety pharmacology was also reported [8–10].

Other sections in the drug development process that can be outsourced include, high throughput screening (HTS), clinical CRO sites, clinical central labs for safety, regulated bioanalysis assays (PK), specialty laboratory for clinical biomarker analysis, manufacturing, marketing, and Information technologies (IT). Another noteworthy interest includes outsourcing overseas to China and India; however, it will be out of the scope for this chapter. This chapter will focus on the strategy and attainment of outsourcing clinical biomarker assays to specialty laboratories. National Institutes of Health (NIH) working group defined biomarkers as "a characteristic that is objectively measured and evaluated as an indicator of normal biological processes, pathogenic processes, or pharmacologic response to a therapeutic intervention" [11]. Biomarkers are a key in drug development as they can be used to depict safety, define mechanisms of action, predict pharmacodynamics, and help in the decision making to stop drug development in early phases leading to cost savings. Biomarker identification, discovery, and assay development are considered core competencies. These core competencies are an ideal fit in this outsourcing paradigm and could be implemented through assay development and clinical sample analysis. The three key steps in successful outsourcing of clinical biomarker analysis, assessment and selection of CRO, application of the strategic partnered relationship model, and trust and execution will be discussed in this chapter.

Assessment and Selection

According to the Tufts Center for the Study of Drug Development (CSDD) there are over 3000 CROs in the US alone, which is a fourfold increase since 2000 [12]. It is quite a task to identify and select the appropriate CRO to meet your assay and outsourcing needs. Therefore, selecting the right CRO may be just as critical as choosing the precise target for drug discovery. A CRO that provides multiple platforms and capabilities ranging from immunoassays, mass spectrometry, genomics, flow cytometry, and molecular pathology, would be ideal as a "one stop shop" as it can be used to help with logistics, costs, and single point of contact. However, this type of CROs is rare and if available, it may not be the industry leaders in every platform. It is critical to identify a CRO that will fit the requirements for the study assay. The four C's; capability, capacity, coordination, and certifications can be used to assess specific CROs.

Selecting a CRO that is aligned with the same capability of the pharmaceutical company such as technology platform will help ensure continuity and reproducibility of the assay results. For example, many CROs that conduct immunoassays will run the meso scale discovery (MSD) platform, however, if the sponsor company developed their assays on the Luminex platform then this CRO will not be compatible with the needs. Other capabilities include the type of equipment that the technologies can use. For example, if the Sponsor is interested in developing clinical flow cytometry assays on the FACSCanto II and the CRO is only operating the FACSCalibur or the FACSCanto I, this instrument incompatibility could delay study start due to errors in detection and may require the redevelopment and optimization of modified monoclonal antibody panels specific to the CRO instrument configuration. The CROs should also have similar equipment and platforms and the redundancy in their equipment to support the assays. In the event that the primary platform equipment goes down or needs servicing, or is dedicated to a single project, equipment redundancy may reduce the risk of variability. Besides recognizing the right platform and equipment it is critical to ensure the CRO's experience, history, training, skill set, personnel, and expertise to run the required assays. Through leveraging the CRO's expertise, new technologies, and assistance in troubleshooting can be established. For example, many CROs are able to run

Sanger sequencing and qPCR; however, the field is moving toward next-generation sequencing (NGS). Thus, the preferred CRO is one proficient in the field of sequencing and able to adapt the new service platforms. The final assessment that should be considered is whether the CRO is able to provide services that cannot be performed internally with higher efficiency, reduced risk, lower cost, and reproducibility. In order to narrow down the CRO searches, the criteria of the first "C" can be discussed and assessed through a capabilities presentation. Another notable step is meeting the CRO for the capabilities presentation face to face. This meetings will allow an open discussion and better understanding of both party's needs. It should be noted that since there may be proprietary and core expertise shared between the sponsor and the CRO, a confidentiality disclosure agreement or an equivalent legal document should be signed and agreed upon by both parties prior to any discussions.

Once the CRO is able to support the company's needs scientifically, the next consideration is to assess its capacity. The company's capacity is defined as the available resources to support the outsourced project such as financial solvency, physical size to accommodate personnel, equipment, storage, and processing, and portfolio. There are several dynamics that should be considered in selecting global versus local CROs such as study size, study complexity, site number, shipping sample logistics, and site location (i.e., domestic, local, region, country, Europe, Asia). The optimal solution is to have a local CRO with a global footprint. This solution will allow a domestic technology transfer, an ease of oversight, and when needed the CRO is able to transfer the protocol and technology through their standardized process internally. Additionally, owing to the stringent regulatory requirement of exporting samples in China, having a CRO with subsidiaries will allow accessibility and collaboration. Another consideration for capacity is whether the CRO is able to handle the study sample size and the sample management logistics. For example, a phase III trial may generate thousands of samples with multiple sites. The CRO should be able to manage the shipments and the large quantity of sample analyses. Lastly, it is imperative to assess the CRO's commitment, customers, and the size of the studies they are supporting. It will be useful to recognize how the CRO prioritize projects between various customers.

The next assessment that should be considered when selecting a CRO is coordination. Coordination is defined as the CRO's ability in managing details and logistics of the company's assay and having an established infrastructure to ensure quality, reproducibility, and error prevention. A CRO should be deemed as an extension of the sponsor lab when outsourcing clinical biomarker assays. Other activities that are often carried out through the sponsor includes, sample management, quality systems, project management, data management, and transfer, and access to similar reagents and controls. Established processes and standard operating procedures (SOP) are required in order to prevent errors and ensure consistencies within the individual CRO site and other intercompany multiple sites. Additionally, when a CRO is selected because of their global footprint, they should have established standardized SOP where the assays can be transferred seamlessly through multiple sites domestically or globally with minimal oversight from the pharmaceutical company. It is also critical that the CRO is able to handle and coordinate a large shipment of samples that may come from multiple clinical sites at multiple time points.

The last assessment when selecting a CRO is certifications. Drug development, regulatory submissions, and other regulatory considerations are noteworthy when working with clinical samples. Outsourcing can be used to share the burden and risk of drug development. The CRO should share the regulatory responsibilities and it is recommended that they operate under controlled environments such as CAP/CLIA, GMP, GLP, or other related certification. Currently, FDA has not yet released their guidance on clinical biomarker assays regulation, thus it is suggested that biomarker assays to be conducted under the same regulatory environment as the approved clinical tests in a Clinical Laboratory Improvement Amendments (CLIA) certified lab [13]. Based on the "fit for purpose" validation guidance, the level of assessment should be based on the intended use of the data [14]. However, it is a best practice that clinical biomarker assays that are intended for prospective patient enrollment, patient segmentation, and diagnostic-related assays should be run in a CAP/CLIA laboratory. Lastly, selecting a CRO that is able to run regulated analyses or the necessities to participate in co-filing with the FDA should be considered.

Strategic Partnered Relationship Outsourcing Model

A recent review highlighted the significance of shifting CRO mindset into a collaboration partnership in the drug development process. Pharmaceutical industry needs to think "outside the box" and support the thought of an open collaboration and knowledge sharing. Additionally, academic collaboration may foster creativity and freedom of discussion. Historically, pharmaceutical industries are more collaborative with academia when compared to CROs. An open and transparent interaction between pharmaceutical industry and CRO may lead to new innovation and increased flexibility through leveraging the expertise of both companies. Proprietary concerns over IP in the past have created gaps in relationships that suppressed innovation and efficiency was noted [15]. The optimal strategy for outsourcing clinical biomarker assays may be attained through a modification of the preferred provider relationship strategy as discussed earlier and the collaboration and knowledge sharing [5, 15]. This new model of "strategic partnered relationship" outsourcing generates a new process that can support the drug development process. Strategic partnered relationship outsourcing requires commitment and builds upon the relationship with the CRO. This model allows the core competencies to be outsourced which lead to higher efficiency while decreasing costs.

The significance of establishing the four C's and relationship are further highlighted since the pharmaceutical industry has carried a majority of the risks. A sizeable amount of work, expectations, and trust are necessary in establishing the new CRO model. Additionally, the culture and attitude toward working with a CRO should be addressed as in the nomenclature of "fee for service vendor" or "outsourcing lab" is redefined as external or collaboration partner. Successful partnerships with CROs in developing biomarker assays may provide early kills on poor pipeline molecules, develop biomarkers to support clinical trial objectives, and help support the development of new molecular entities. A recent example of the strategic partnered relationship model was demonstrated by Astra Zeneca's partnership with a CRO. As a result of this relationship a reduction in management and oversight, an increase productivity, and reduce risk were observed [5, 7]. In order for the company to meet their needs, they may need to work with several CROs over the course of a project or several projects to find the best fit.

As mentioned previously, outsourcing clinical biomarker assays and selection will be based on the platform and capabilities of the CRO in conducting the necessary assay development or sample analyses. It is also important to conduct a site visit shortly after the capabilities presentation as stated previously. The time and resource commitment to investigate the site and recognize the partner's operations, facility, systems, and personnel demonstrates the desire in developing the relationship. During the site visit, an open discussion in long term relationships, shared knowledge, and shared core competencies will also foster the relationship. Other key factors that may help form the relationship include sharing key technologies, platforms, or best practices. For example, understanding the type of automation the sponsor is utilizing or having the knowledge of cutting edge NGS platform in the field. In order to reinforce the relationship, expectations should be established early on and discussed often. As part of the expectations, the following factors should also be discussed; co-developing a communication plan, integrating external, and internal teams through weekly meetings, sharing data, and seeking feedback on data analysis from the partner.

Trust and Execution

Establishing trust is critical in order to continue the relationship success. To establish trust, the external partner should be considered as an extension of the sponsor laboratory and be expected to operate in the same or higher level. The external partner often has the industry redundancy including sample capabilities, equipment, platform, and sometimes personnel. Due to these redundancies, they are also considered experts in the field and their expertise should be leveraged and allowed to drive the process and development of science. Taking a step back and "releasing the reins" in a move toward collaboration will foster trust. Auditing either by the sponsor or a third party can provide evidence that the external partners are compliant with regulatory guidance and operating at an acceptable level to the sponsor. Audits are typically requested and performed by the sponsor prior to transferring any work to an external partner. Additionally, adhering to sponsor and regulatory compliance will help decrease any risks associated with the outsourced project.

Another key factor in establishing relationship and trust is communication. Discussing, understanding and co-developing the external partner's issue escalation and communication process will prevent frustrations, timelines delay, or failed projects in the future. For example, a recent review noted the triumph of Johnson & Johnson in outsourcing a core competency, compound collection in drug discovery through implementation of an open collaboration, transparency, and visibility. This process was initially thought of as the "crown jewels" in drug discovery, however was successfully outsourced based on the new model [16]. In clinical biomarker assay development, many of the issues can be prevented if the sponsor is included to participate in troubleshooting activities in the early stages. However, this relationship can only be established if the communication and issue escalation plan have been agreed upon and executed. When developing an open communication plan, the expectation of flexibility on the external partner should be discussed early on. Due to varying timelines, accelerated objectives and filings, and the need for data to support decision making, having a flexible external partner will prevent frustration. Lastly, the biggest constituent of trust is the execution and awarding the external partner with the project. However, it is important to note that the sponsor does not remove all oversight and communication from the assay once the project is outsourced. It is the sponsor's responsibility to provide oversight, resources, support troubleshooting, and ongoing requirements to the external partner. Ongoing external partner monitoring, follow-up via recurrent meetings, and continued site visits are required in order to continue a successful relationship. The completion of successful outsourcing will establish a relationship that can be fruitful and rewarding for future projects and help support an innovative drug development process.

Discussion

The need to develop NME and NBE through cost cutting, increased efficiency, and decreased timelines in the pharmaceutical industry have pressed many core competencies and phases of the drug development process to be outsourced to CROs. The recent increase in the number of CRO's available to support the industry has made drug development more efficient, able to provide technical expertise, and open up resources within the company to develop more work discoveries. The perception of outsourcing has changed from the tactical model to a strategic partnered relationship model which includes CROs as an integrated partner with knowledge sharing. These newly developed relationships may spark innovations in drug discovery and provide best practices for new cutting edge platforms and capabilities. Clinical biomarker assay strategic outsourcing can be a key constituent in the drug discovery process and support the transformation in the industry. However, there are a number of lessons learned and common pitfalls that should be avoided. One such pitfall is outsourcing major core competencies that may pose a risk of losing the edge of clinical success in addition to the loss of innovation [6]. As mentioned above, companies are generally concerned with IP loss; however, conducting business in a closed protective knowledge sharing will diminish innovation and success. A balance between knowledge sharing and proprietary information maintenance should be carried out. The strategy that is practiced commonly within the industry is to keep the major core competencies such as assay development in house, and outsource certain core competencies such as sample analysis to the external partner.

It is noteworthy to have a well-defined outsourcing plan to include the following, issue escalation and communication plan, points of contact identification between the CRO and sponsor, statement of work contract, and allocated budget. A poor or undefined plan will result in a risk of failed outsourcing project and constant rotation of external partners. This illustration may create long lead time for outsourced projects and wasted time in evaluating external partners. Lastly, it is important to have a qualification process in place that documents and accounts for all the information gathered in the assessment section. In summary, the CROs should be selected utilizing the four C's, and once the CROs meet the requirements established by the sponsor, they can be considered qualified and approved providers. Consequently future-outsourcing projects can be contracted to the preferred providers based on their capabilities. Outsourcing can be proven beneficial if one is able to balance the success criteria described above with the risk assessment while avoiding pitfalls.

•	The movement of outsourcing certain core competencies transformed the outsourcing strategy from a tactical outsourcing model to strategic partnered relationship
•	The new strategic partnered relationship model outsources facets of the drug development process based on collaboration, open knowledge sharing, and relationship development
•	Biomarkers are a key in drug development as they can be used to depict safety, define mechanisms of action, predict pharmacodynamics, and help in the decision making to stop drug development in early phases leading to cost savings. Biomarker identification, discovery, and assay development are considered core competencies
•	There are three key steps in successful outsourcing of clinical biomarker assays; assessment and selection of CRO, application of the strategic partnered relationship outsourcing model, and trust and execution
•	The four C's; Capability, Capacity, Coordination, and Certifications can be used to assess specific CROs

References

1. Kaitin KI (2010) The lanscape for pharmacuetical innovation: drivers of cost-effective clinical research. Pharm Outsourcing
2. Kaitin KI, DiMasi JA (2011) Pharmaceutical innovation in the 21st century: new drug approvals in the first decade. Clin Pharm Ther 89(2):183–188
3. DeRuiter J, Holston PL (2012) Drug patent expirations and the "Patent Cliff". US Pharm 37(6):12–20
4. Tufts Center for the Study of Drug Development (2006) CRO contribution to drug development is substantial and growing globally. Impact Report 8(1):1–4

5. Jones J, Minor M (2010) New strategic outsourcing models to meet changing clinical development needs. Perspect Clin Res 1(2):76–79
6. Klopack TG (2000) Strategic outsourcing; balancing the risks and the benefits. Drug Discov Today 5(4):157–160
7. Clark DE (2011) Outsourcing lead optimization—the eye of the storm. Drug Discov Today 16(3/4):147–157
8. Schwardt O, Kolb H, Ernst B (2003) Drug discovery today. Curr Top Med Chem 3:1–9
9. Meerpoel L, Schroven M, Goris K, Demoen K, Siobhan M (2006) Outsourcing to exploit a key asset. Drug Discov Today 11(11/12):556–560
10. Ewart L, Gallacher DJ, Gintant G, Guillon JM, Leishman D, Levesque P, McMahon N, Mylecraine L, Sanders M, Suter W, Wallis R, Valentin JP (2012) How do the top 12 pharmaceutical companies operate safety pharmacology? J Pharm Tox Methods 66:66–70
11. Biomarkers Definition Working Group (2001) Biomarkers and surrogate endpoints; preferred definitions and conceptual framework. Clin Pharmacol Ther 69:89–95
12. Gertz K, Lamberti MJ, Mathias A, Stergiopoulos S (2012) Resizing the global contract R&D services market. Contract Pharma
13. Lykins R, Kadam S, Leppert C, Pascarella-Hallett A (2011) Outsourcing clinical biomarker development; a pragmatic approach. J Pharm Biopharm Contract Serv
14. Lee JW, Devanarayan V, Chen Barrett Y, Weiner R, Allinson J, Fountain S, Keller S, Weinryh I, Green M, Duan L, Rogers JA, Millham R, O'Brien PJ, Sailstad J, Khan M, Ray C, Wagner J (2006) Fit-for-purpose method development and validation for successful biomarker measurement. Pharm Res 23(2):312–328
15. Robertson GM, Mayr LM (2011) Collaboration versus outsourcing; the need to think outside the box. Future Med Chem 3(16):1995–2020
16. Ramsey DL, Kwasnoski JD, Caldwell GW (2012) Small molecule compound logistics outsourcing—going beyond the "thought experiment". Curr Top Med Chem 12:1298–1308

Author Biography

Jeffrey A. Tsou currently manages RUO and IVD programs in Reproductive and Genetic Health at Illumina. Previously, he was the Head of Global Outsourcing and Operations in the Molecular Biomarkers and Diagnostics Department at Merck located in Rahway, New Jersey. His focus was on establishing the outsourcing process and vendor oversight for all biomarker assay and biomarker sample analyses that are placed at CROs.

In Situ Hybridization in Clinical Biomarker Development

Usha Singh, Marisa Dolled-Filhart and Dianna Wu

Abstract Biomarkers are being utilized from early to clinical phases of drug development. Nucleic acid-based biomarkers have proven their significance in predicting disease occurrence, identifying the reason for variable drug responses and helping in personalized targeted therapy. There are various techniques available for analyzing nucleic acid-based biomarkers in body fluids and in tissue homogenates; however, detection of tissue-based biomarkers in the context of tissue/cell morphology is still limited to the use of in situ hybridization (ISH) technology. This book chapter provides an overview of available ISH technologies, efforts made toward the advancement of ISH to increase the sensitivity and specificity of assays and their applications in clinical biomarker.

Key terms

Biomarker	Biomarker is an indicator of normal and pathogenic process, progress of disease, or effect of therapeutic interventions and it can be measured in body fluids or tissues
In situ *hybridization* (ISH)	In situ hybridization (ISH) is a type of hybridization method that uses a labeled complementary DNA or RNA strand to localize a specific DNA or RNA sequence in a section of tissue
Formalin fixed paraffin embedded (FFPE) tissue	It is tissue processing method used in histology labs to preserve and archive clinical specimens
Tissue Micro Array (TMA)	TMA consists of multiple separate tissue cores that are assembled in array fashion to allow multiplex histological analysis

U. Singh (✉) · M. Dolled-Filhart · D. Wu
Translational Medicine, Merck & Co., 126 East Lincoln Ave, Rahway, NJ 07065, USA
e-mail: usha.singh@merck.com

M. Dolled-Filhart
e-mail: marisa.dolled-filhart@merck.com

R. Weiner and M. Kelley (eds.), *Translating Molecular Biomarkers into Clinical Assays*, AAPS Advances in the Pharmaceutical Sciences Series 21,
DOI 10.1007/978-3-319-40793-7_18

Introduction

The cost burden of research and development in drug industries is rising every year while the number of new drug approvals continues to decline. The attrition rate for developing a drug is reported to be 10,000:1, only 1 in 10,000 new drug candidates that enter in preclinical testing makes to the market [1]. This high attrition rate is motivating pharmaceutical industries to modify traditional drug testing paradigms and focus on integration of translational biomarkers. Effective implementation of biomarkers to understand the disease state, prognosis, and response to therapy is a promising approach to identify problems in early stages of drug development.

Biomarkers (biological markers) are indicators of a normal or pathogenic process, progress of disease, or effect of therapeutic interventions. Biomarkers are being utilized from early drug discovery to development to understand the cellular, biochemical, or molecular alterations and their relationships with disease or treatment. In early stages of drug development, biomarkers are used for developing animal models, proof of concept studies, molecular profiling, bridging animal and human pharmacology, and evaluating safety studies. In late drug development, biomarkers can be utilized in the evaluation of dose selection, patient stratification, risk identification, and evaluating the response to therapy. Personalized medicine is the customized treatment for individual patient on the basis of patient's genetic and physiological profile.

Recent advances in proteomics and genomic technology have discovered sensitive and specific biomarkers leading to improved diagnosis. In combination with the advancement of molecular biology techniques, biomarkers will be a dynamic and powerful tool to understand the spectrum of pathogenic conditions, disease progression, therapy selection, and possibly a new direction in the clinical trial design.

Types of Biomarkers

Biomarkers are mainly used for safety or efficacy of a drug candidate and can be divided into two categories: safety biomarkers and efficacy biomarkers. In addition to physical examination and vital signs, safety biomarkers should be measured to detect the early indications of specific toxicity. As liver is the first resort of drugs, examining drug-induced liver injury is the critical step for drugs toxicity evaluation. Renal safety biomarkers come next as the kidney is a crucial organ which eliminates undesirable end products and toxicants. Other safety biomarkers are hematology, bone, metabolic and other serum specific biomarkers.

Efficacy biomarkers are indicators of disease conditions and they can be divided into four groups: diagnostic (screening), prognostic, predictive, and pharmacodynamic biomarkers. Diagnostic biomarkers are used to differentiate the pathogenic versus normal state. Prognostic biomarkers predict the outcome or progression of disease. Predictive biomarkers select patients for specific treatment or to predict the response to a specific therapy. Pharmacodynamic biomarkers are used to determine the response of drug and selection of dosage. Biomarkers can also be classified based on their characteristics, such as imaging biomarkers (CT, PET, MRI) or molecular biomarkers. Molecular biomarkers have biophysical properties and can be measured in biological samples, such as human tissues, cells, and body fluids.

Importance and Applications of ISH

Genomic research has proven the potential of nucleic acid-based biomarkers to predict disease occurrence, identify the reason for variable drug responses, and help in personalized targeted therapy [2]. There are several techniques available for measuring biomarkers in body fluids and in tissue homogenates; however, detection of tissue-based biomarkers in the context of tissue/cell morphology is still limited to the use in situ hybridization (ISH) technology. ISH has been used widely to localize the expressions of nucleic acids (DNA, RNA, and miRNA) at the cellular level in a heterogeneous cell population [3–5].

In infectious disease, many DNA probes are available for the identification of foreign genes, including bacteria, viruses, and fungi in tissue sections [6], which provides information about the etiology of the infectious disease. ISH is also useful for chromosomal gene mapping, characterization of genetic abbreviations and identification of genetic abnormalities [7]. These findings can help in the diagnosis of a genetic disease and prognostic outcomes of disease. Detection of heterogeneity of gene expression and variability in genetic profiling among individuals are the key factors for various responses to the same treatment and needs to be considered for successful clinical drug testing. Detection of genes encoding cell structural proteins, growth factors and their receptors including tumor-associated markers, represent potential areas of application of ISH methods for clinical research.

ISH Technologies and Challenges

ISH is the only technology that allows the cellular localization of the genetic sequence of interest to be studied [8]. ISH method requires a labeled nucleic acid probe to detect and localize specific RNA or DNA sequences in a tissue. The probe can either be radioactively labeled, fluorescently labeled (abbreviated FISH) or enzyme labeled for chromogenic detection (CISH). Today the FISH and CISH detection methods are preferred as they are safer, faster, and produce high resolution images compared to radioactive ISH. CISH has the ability to view the signal and tissue morphology simultaneously and slides can be stored for long time. FISH is a more useful technique for gene amplification, deletion, and translocation analysis and multiple targets can be detected in the same sample.

Based on the nature of detection molecule, in situ hybridization methods can be divided into three types, (1) DNA ISH, (2) RNA ISH and (3) microRNA (miRNA) ISH.

DNA ISH: DNA ISH can be used to determine the structure of chromosomes. DNA is relatively stable and the assay can be performed easily compared to RNA ISH. For DNA ISH, the probes can be double-stranded DNA, single-stranded DNA (ss-DNA) or synthetic oligonucleotides.

RNA ISH: The RNA ISH is used to localize the RNAs (mRNA, long noncoding RNA or IncRNA and miRNA) within tissue sections or cells. For RNA ISH, the probes can be riboprobe, DNA or synthetic oligonucleotides. Developing RNA-based assays can be challenging since there is a need to avoid the contamination with RNAse.

miRNA ISH: MicroRNAs (miRNAs) are small, $\sim$20 nucleotides noncoding RNAs that bind to messenger RNA transcripts, and alter the protein expression. Changes in the expression levels of specific miRNAs have been associated with a variety of disease conditions [9–11]. Several miRNA ISH protocols are available for a variety of specimens [12–16]. Detection of miRNA is more challenging as they are smaller in size and significant miRNAs can be lost from tissues during processing. Longer fixation of tissues is recommended for better preservation of miRNA during ISH [17].

There has been considerable development in ISH in last few decades and it is becoming a popular molecular diagnostic tool in clinical laboratories. The advancement in ISH has been achieved mainly by: (1) increasing the sensitivity by amplifying signal, (2) designing stable probes, (3) developing assays for FFPE tissues for better morphology, (4) multiplexing the assays to detect more than one target, (5) developing automated and robust methods and (6) to be able to quantify the signal.

There are new more sensitive technologies available to detect RNAs by in ISH such as RNAscope, Quantigene ViewRNA, and branched DNA ISH [18–20]. These technologies use unique probe design and signal amplification to achieve single RNA molecule visualization in samples (cells, tissues, and CTCs). RNAscope ISH can be performed on frozen or formalin fixed paraffin embedded (FFPE) tissues either by using chromogenic dyes or fluorescent dyes. The method uses unique double Z-probes, where two independent probes (a pair of double Z) are required to hybridize to the target sequence for signal amplification to occur. The initial steps are similar as other ISH procedure which involves the fixation and permeabilization by enzymatic treatment. The probes (Z) are then hybridized in pairs (ZZ) to the target sequence. A preApmplifier (PreAMP) molecule hybridizes to each Z probe pair, and then multiple Amplifier (AMP) molecules hybridize to each PreAMP. Finally, multiple HRP-labeled probes hybridized to each AMP and 3, 3′-diaminobenzidine (DAB) substrate is used for colorimetric detection of target RNA. The assay can be multiplexed and performed either manual or automated on a Ventana platform (Ventana Medical Systems, Tuscan AZ) [18]. There are other technologies, e.g., branched DNA ISH, and QuantiGene ViewRNA assay [19, 20], use more or less same approach of probe designing to enhance the specificity and sensitivity of ISH.

Most of the available ISH procedures are done manually; however, ISH is a lengthy and tedious process. The manual staining procedure needs extra steps which can introduce more inter- and intra assay variability. There are various instruments available to perform automated ISH procedures and undoubtedly automation is helpful in relieving laboratory personnel of repetitive actions and in improving the reproducibility of results. Some of the automated stainers available are Ventana Ultra^{TM}, Discovery XT^{TM}, DAKO, Lab Vision Autostainer 360^{TM}, and intelliPath-FLX^{TM} to perform IHC/ISH. All of these instruments are primarily designed for IHC however most of them are also capable of performing ISH. For automated ISH, the instrument must be able to apply controlled heat to slides for denaturing the DNA. Manual ISH requires hybridization chambers or oven (RNAscope) to perform denaturation and hybridization process. The choice of instrument is evaluated on the basis of its use, complexity, and cost involved with assay development. Usually, there are two main criteria in selecting the instrument, the functional capabilities, and the quantity of slides that can be processed in one run. Some of the platforms are fully automated while others needs few initial steps to be performed offline These automated systems also save protocols which can be accessible later and print run reports for records and further confirmation of run accuracy. Automation is very critical for clinical labs for obtaining consistent results, managing data, and developing assays for in vitro diagnostics. There are several IVD ISH assays available such as PathVysion FISH Kit, HER2 CISH PharmDx kit, and SPOT-Light HER2 CISH Kit to detect amplification of the HER-2/neu gene, the Vysis ALK Break Apart FISH Probe Kit to detect the ALK gene [21, 22]. Although automation is an important step to move the technology forward, these instruments are costly and have limited flexibility in terms of reagents and protocol design.

Assay Development

Selection of Assay: Selection of ISH assay depends on the expression level of target sequence, tissue type, species, available probes and resources to develop the assay.

Probes: There are different types of probes, such as double-stranded DNA probes, single-stranded DNA probes, RNA probes, and synthetic oligonucleotide probes. The selection of probe depends on the application of ISH and desired sensitivity of the assay. Double-stranded DNA probes are less sensitive as they have tendency to hybridize with each other thus fewer probes are available to bind with target sequence. The RNA probes are sensitive to degradation by RNAse and therefore need an RNAse free working environment before hybridization; however, after hybridization the RNA–RNA interaction is resistant to RNAses. Oligonucleotides probes are synthetically produced and relatively smaller in size (20–40 base pairs). These probes can easily migrate through the cell membranes. They are relatively less sensitive than DNA and RNA probes. Since direct detection approaches using oligo probes may not be sufficiently sensitive for detection, most of the new technologies use preamplifier sequences which are labeled with several hapten molecules as described earlier in the 'Technologies' section. Another modified RNA nucleotide probe is a locked nucleic acid (LNA) probes in which the ribose ring is locked into C3′-endo conformation by a 2′-O, 4′-C methylene bridge [12]. LNA probes are resistant to exo- and endonucleases which provides high thermal stability. These probes are more suitable for the detection of short RNA and DNA targets such as miRNA. The LNA probes have been utilized successfully in conducting microarray profiling, PCR, and ISH [12, 13, 17, 23]. The automated ISH methods for single and dual labeling of miRNA have been developed successfully [17, 24]. Probes are either labeled with radioisotopes (32P, 35S and 3H), biotin, digoxigenin (chromogenic detection) or fluorochromes. Singleplex assays mostly prefer chromogenic labeling while multiplex assays require the fluorescence labeled probes.

Sample collection and preparation: Hospitals and labs have standard operating procedures (SOP) to collect tissue samples. After tissue collection the samples are either processed in-house or shipped to another lab for processing and sample analysis. During study planning, instructions are provided for fixation, processing, shipping and handling of the clinical samples. Factors that may influence the final ISH results in clinical FFPE samples include (a) surgery method and time (b) time and storage temperature after resection until processing into formalin (c) formalin fixation time and temperature and (d) shipping and handling procedure. Common fixatives for tissues are either cross linking fixatives (e.g., formaldehyde, glutaraldehyde) or precipitating fixatives (ethanol, methanol). For nucleic acids preservation precipitating fixatives are better however, they cause shrinkage and hardening of tissues which is why they are not commonly used. FFPE tissues are usually fixed for 24–48 h in 10 % neutral buffered formalin (NBF) after collection. It has been observed that for ISH longer fixation preserves RNA better in tissues [17]. Frozen sections need to be fixed for 20 min to 1 h in any fixatives (crosslinking or precipitating) depends on the assay, to preserve the nucleic acids and cellular morphology.

Staining procedure: The main steps in ISH procedure involves deparaffinization/enzyme treatment, blocking, prehybridization, denaturation, hybridization, and visualization of the probes on target tissues. FFPE tissue sections require deparaffinization to unmask the epitope either by heat or enzymatic treatment. The choice of blocking depends on the amplification or enzymatic detection methods used in the assay. Blocking of tissue sections before hybridization prevents false positive signal generated from endogenous biotin or enzymes during detection. Prehybridization step is recommended to prevent background staining in hybridization with RNA as target and to maintain the hybridization stringency. Denaturation of tissues is necessary when target sequence is DNA and this can be achieved by heat or higher salt solution or mixture of both. The hybridization depends on the design of probe and hybridization conditions. The stringency of washes after hybridization is an important step to reduce the nonspecific binding of probes. The hybridization conditions must be optimized for each assay according to the nucleic acid sequence, Tm value, probe design, and platform of ISH assay.

The signal of bound probe can be detected by autoradiograph (radioactive ISH) or by immunolableing (FISH and CISH). In the immunolabeling method, the hybridized probe is either labeled with fluorochrome, enabling direct detection or a hapten (i.e., biotin, digoxigenin, DNP) that can be detected indirectly [7]. In indirect detection method, the hapten can be identified by specific antibodies which are labeled either with a fluorescein or an enzyme (*Horseradish peroxidase*, *Alkaline phosphatase*). The fluorescein signal can be visualized by fluorescence microscope. The enzyme labeled probes need substrate (Diaminobenzidine tetrachloride or Nitroblue Tetrazolium) to generate chromogenic signal and they can be visualized by light microscope. New technologies such as RNAscope for mRNA detection or automated miRNA ISH protocols using LNA probes [17], uses automated protocols which are straight forward and needs less troubleshooting.

Assay optimization and validation: Assay optimization processes include the selection and optimization of reagents, control tissues, procedure, detection methods, and instrumentation platform. ISH assays are most often developed on archived human samples available in tissue banks which to a wide extent are FFPE tissue specimens. Since these archived tissues might be fixed and processed differently, it is recommended to test several tissues to optimize the assay. Positive control probes reveal an intense ISH signal and diluting the probe concentration to the level of sensitivity may be helpful to test the quality of individual tissue sample. Various other controls can be used to check the specificity of assay such as omission of specific probes in hybridization reaction, isotype control for antibody, negative control probe, negative control tissue, and combining ISH with IHC to localize the translated protein in the same cell type.

Assay validation is required to comply with the standards for clinical biomarker assays. There is no defined procedure to validate ISH assay. The fit for purpose validation procedure includes testing of sensitivity, normal cutoff, positive detection reference range, reproducibility, specificity, and stability. Sensitivity of assay can be measured on a range of positive control tissues which have low to high level of gene expression. Usually a tissue microarray (TMA) containing multiple tissue samples are useful for assay optimization. In clinical assay development, validation includes few extra steps such as proficiency testing, assessment of employee competency, instrument variability; inter-lab variation and correlation between clinical findings.

Sample Analysis

Usually the assay is developed and validated in advance of clinical sample analysis using appropriate control samples. Hematoxylin and eosin (H&E) staining is performed to test the morphology and clinical diagnosis of tissues. Serial sections are used for ISH assays according to the developed procedure. For quality control and detection range, a slide of TMA (with low to high range of tissues) is included in each staining run. Appropriate positive and negative control slides are also included with samples in each run to check the performance of assay.

Result Interpretation

Qualitative analysis is used to define the subcellular localization of nucleic acids in tissues and semi quantitatively to compare the expression between normal and abnormal cells/tissues. The analysis must always be done with the understanding of tissue type being tested, probe(s) used and the expected signal pattern of normal versus abnormal cell type. A tissue section may have more than one type of abnormal cells with characteristic morphology. The semi-quantitative method is manual scoring which is usually performed by counting the mean signal (dots) per nuclei (cell). At least two people should score a substantial number of cells in each study. These people should be unaware of each other's scores. The signal pattern of normal population is analyzed by testing several normal tissues during assay development.

The large effort has been made to develop the automated cell imaging, segmentation, and scoring systems with multichannel capabilities using FISH [25–27]. Few of the available image analysis tools are Definiens RNAscope SpotStudio™ (Definiens AG, Munich, Germany), Metafter™ system (MetaSystems Group, Inc., MA) [25] and Duet™ (BioView, MA) automated system [26] for examining numerical chromosomal aberrations in various types of cells. Automated image analysis for ISH should be further explored as it requires appropriate scanners, reliable, and validated algorithms and analysis methods to conclude the data. In clinical biomarker analysis where biomarkers are used for patient selection, it is critical to demonstrate the statistical significant relevance with regards to clinical variables.

Future Perspective of ISH

The future perspective of new technologies for ISH will make it a very powerful tool for biomarker analysis. With advancement of genomic research, where most of the genes are identified by "grind and bind" tube analysis, ISH is the only technique to localize the nucleic acid in content of tissues. Extensive efforts have been made toward the advancement of ISH includes, use of tissue microarray, ISH methods for FFPE tissues, different types of probes with amplification detection systems to increase sensitivity and specificity of assay, and automated platforms to perform ISH. Computer-based image analysis quantification methods have also added significant strength to this technology for providing more accurate and reproducible results, which is critical for diagnostic assays. Multiplexing ISH to detect more than one nucleic acid or nucleic acid detection with protein detection has various advantages to identify the role of biomarkers in normal and pathophysiological conditions. For diagnostic purposes, ISH needs the comprehensive validation evaluation. There is wide variability in clinical setting which is why guidance on parameters relating to tissue fixation, processing, ISH protocols, and evaluation of results needs to be streamlined for potential clinical use of ISH.

Summary

- This book chapter is a reference for scientists who are interested in developing ISH-based biomarker assays.
- It is intended to provide an overview of different available technologies for ISH, recent advancement, platforms, and process of clinical biomarker assay development.
- This chapter also focuses on some pros and cons of technological advancements and challenges we face to develop clinical biomarker assays.

References

1. Halim AB (2011) Biomarkers in drug development: a useful tool but discrepant results may have a major impact. In: Kapetanović I (ed) Drug discovery and development—present and future. ISBN: 978-953-307-615-7. doi:10.5772/27526 (InTech, Chapter 18)
2. Frank R, Hargreaves R (2003) Clinical biomarkers in drug discovery and development. Nat Rev Drug Discov 2:566–580
3. Schwarzacher T (2003) DNA, chromosomes, and in situ hybridization. Genome 46(6): 953–962
4. Jin L, Chandler WF, Smart JB, England BG, Lloyd RV (1993) Differentiation of human pituitary adenomas determines the pattern of chromogranin/secretogranin messenger ribonucleic acid expression. J Clin Endocrinol Metab 76(3):728–735
5. Varma VA, Cerjan CM, Abbot KL, Hunter SB (1994) Non-isotopic in situ hybridization method for mitochondria in oncocytes. Histochem Cytochem 42:273–276
6. Kiflemariam S, Andersson S, Asplund A et al (2012) In situ hybridization on tissue arrays for validation of novel cancer and tissue-specific biomarkers. PLoS ONE 7(3):e32927
7. Bishop R (2010) Applications of fluorescence in situ hybridization (FISH) in detecting genetic aberrations of medical significance. Biosci Horiz 3(1):85–95
8. Polak JM, O'D McGee J (1990) In situ hybridization: principles and practice. Oxford University Press, Oxford, pp 10–30
9. Hanke M, Hoefig K, Merz H et al (2010) A robust methodology to study urine microRNA as tumor marker: microRNA-126 and microRNA-182 are related to urinary bladder cancer. Urol Oncol 28(6):655–661
10. Xie Y, Todd NW, Liu Z et al (2010) Altered miRNA expression in sputum for diagnosis of non-small cell lung cancer. Lung Cancer 67:170–176
11. Qian W, Lu Z, Li H et al (2011) Next-generation sequencing of MicroRNAs for breast cancer detection. J Biomed Biotechnol 2011:1–7
12. Jorgensen S, Baker A, Moller S, Nielsen BS (2010) Robust one-day in situ hybridization protocol for detection of microRNA in paraffin samples using LNA probes. Methods 52 (4):375–381
13. Pena JTG, Sohn-Lee C, Rouhanifard SH et al (2009) miRNA in situ hybridization in mammalian tissues fixed with formaldehyde and EDC. Nat Methods 6(2):139–141
14. Nuovo GJ (2008) In situ detection of precursor and mature microRNAs in paraffin embedded, formalin fixed tissues and cell preparations. Methods 44:39–46
15. Silahtaroglu AN, Nolting D, Dyrskjot L et al (2007) Detection of microRNAs in frozen tissue sections by fluorescence in situ hybridization using locked nucleic acid probes and tyramide signal amplification. Nat Protoc 2:2520–2528

16. Liu X, Sempere LF, Ouyang H et al (2010) MicroRNA-31 functions as an oncogenic microRNA in mouse and human lung cancer cells by repressing specific tumor suppressors. J Clin Invest 120(4):1298–1309
17. Singh U, Keirstead N, Wolujczyk A et al (2013) General principles and methods for routine automated microRNA in situ hybridization and double labeling with immunohistochemistry. Biotech Histochem 1–8
18. Wang F, Flanagan J, Su N et al (2012) A novel in situ RNA analysis platform for formalin-fixed paraffin-embedded tissues. J Mol Diagn 14(1):22–29
19. Player NP, Lu-Ping Shen, Kenny D et al (2001) Single-copy gene detection using branched DNA (bDNA) in situ hybridization. J Histochem Cytochem 49(5):603–611
20. Affymetrix QuantiGene®ViewRNAISH (2012) Tissue assay on Leica BOND RX staining platform. Press Release
21. Cayre A, Mishellany F, Lagarde N, Penault-Llorca F et al (2007) Comparison of different commercial kits for HER2 testing in breast cancer: looking for the accurate cutoff for amplification. Breast Cancer Res 9(5):R64
22. Thunnissen Erik, Bubendorf Lukas, Dietel Manfred et al (2012) EML4-ALK testing in non-small cell carcinomas of the lung: a review with recommendations. Virchows Arch 461 (3):245–257
23. Maruo R, Yamada H, Watanabe M et al (2011) mRNA quantification after fluorescence activated cell sorting using locked nucleic acid probes. Mol Biotechnol 49(1):42–47
24. Bartlett JM, Campbell FM, Ibrahim M et al (2011) A UK NEQAS ISH multicenter ring study using the ventana HER2 dual-color ISH assay. Am J Clin Pathol 135(1):157–162
25. Schunck C, Johannes T, Varga D et al (2004) New developments in automated cytogenetic imaging: Unattended scoring of dicentric chromosomes, micronuclei, single cell gel electrophoresis, and fluorescence signals. Cytogenet Genome Res 104:383–389
26. Lev D, Daniely M, Zudik A et al (2005) Automatic scanning of interphase FISH for prenatal diagnosis in uncultured amniocytes. Genet Test 9:41–47
27. Wang YP, Castleman KR (2005) Normalization of multicolor fluorescence in situ hybridization (M-FISH) images for improving color karyotyping. Cytometry A 64A:101–109

Erratum to: Translating Molecular Biomarkers into Clinical Assays

Russell Weiner and Marian Kelley

Erratum to:
R. Weiner and M. Kelley (eds.), *Translating Molecular Biomarkers into Clinical Assays*, AAPS Advances in the Pharmaceutical Sciences Series 21, DOI 10.1007/978-3-319-40793-7

The book was inadvertently published with an incorrect volume number as 37 whereas the correct number for this volume is 21. The erratum book has been Updated with the correction.

The original online version for this book can be found at 10.1007/978-3-319-40793-7

R. Weiner (✉)
Translational Biomarkers and Personalized Medicine, Edison, NJ, USA
e-mail: usha.singh@merck.com

M. Kelley
MKelley Consulting LLC, West Chester, PA, USA
e-mail: mmk48@comcast.net

R. Weiner and M. Kelley (eds.), *Translating Molecular Biomarkers into Clinical Assays*, AAPS Advances in the Pharmaceutical Sciences Series 21,
DOI 10.1007/978-3-319-40793-7_19

Index

R. Weiner and M. Kelley (eds.), *Translating Molecular Biomarkers into Clinical Assays*, AAPS Advances in the Pharmaceutical Sciences Series 21,
DOI 10.1007/978-3-319-40793-7

MIX
Papier aus verantwortungsvollen Quellen
Paper from responsible sources
FSC® C105338

If you have any concerns about our products,
you can contact us on
ProductSafety@springernature.com

In case Publisher is established outside the EU,
the EU authorized representative is:
Springer Nature Customer Service Center GmbH
Europaplatz 3, 69115 Heidelberg, Germany

Printed by Libri Plureos GmbH
in Hamburg, Germany